Carlão Limeira

About the Author

Karen Strier (Ph.D., Harvard University, 1986) is Associate Professor of Anthropology at the University of Wisconsin-Madison. She studied yellow baboons in Kenya before beginning her long-term research on the muriqui monkeys in Brazil.

FACES IN THE FOREST

FACES IN THE FOREST

The Endangered Muriqui Monkeys of Brazil

KAREN B. STRIER

Department of Anthropology
University of Wisconsin-Madison

New York Oxford
OXFORD UNIVERSITY PRESS
1992

Oxford University Press

Oxford New York Toronto
Delhi Bombay Calcutta Madras Karachi
Kuala Lumpur Singapore Hong Kong Tokyo
Nairobi Dar es Salaam Cape Town
Melbourne Auckland

and associated companies in
Berlin Ibadan

Published by Oxford University Press, Inc.,
200 Madison Avenue, New York, New York 10016

Library of Congress Cataloging-in-Publication Data
Strier, Karen B.
Faces in the forest ; the endangered muriqui monkeys of Brazil /
by Karen B. Strier.
p. cm. Includes bibliographical references and index.
ISBN 0-19-506339-2
1. Woolly spider monkey—Brazil, Southeast. 2. Endangered
species—Brazil, Southeast. I. Title.
QL737.P925S77 1992
599.8′2—dc20 91-46767 CIP

9 8 7 6 5 4 3 2 1

Printed in the United States of America
on acid-free paper

To my parents, who gave me dreams

and

to Rich, who has always been there

Foreword

It was in 1974 when, with my colleague, ornithologist Ney Carnevalli, I was lucky to come across a group of muriquis in a remote patch of the devastated Atlantic forest in a coffee farm in the east of the state of Minas Gerais. I was looking for opossums and Ney for tinamous, but we both realized the great importance of our discovery, and the extreme good fortune that the owner of the farm, the remarkable pioneer Feliciano Miguel Abdalla, was also aware of the treasure that he protected. Little did I realize, however, the extent to which this would change my life, and, I would think it reasonable to say, the extent to which it changed an entire Department of Zoology in the Federal University of Minas Gerais where we both worked. Feliciano's muriquis inspired the establishment of a master's course in wildlife management and a Foundation dedicated to the conservation of biological diversity, which today administers the Caratinga Biological Station. Our efforts to provide security and protection for the muriquis and their forest brought us into contact with many international conservationists and primatologists, and directed many of our students towards primate studies and wildlife conservation. Notable was the support and enthusiasm of Russell Mittermeier, who galvanized international interest in the establishment of research projects on the ecology and conservation of the initially extremely elusive muriquis, which today, along with the golden lion tamarins, symbolize the demise of the once vast Atlantic coastal forest of Brazil.

I remember with great affection the first studies by the tireless Akisato Nishimura, who was to introduce Feliciano to a new breed of human being—the primatologist. The eccentricity of spending the entire day searching for the monkeys and eventually observing them for the only too brief moments was a revelation, and Nishimura prepared Feliciano and the muriquis for those who came later. Feliciano was soon to realize that his muriquis excited the interest not only of a lanky, smiling Japanese and an enthusiastic Brazilian professor, but also numerous Brazilians and gringos of all shapes and sizes. Perhaps most surprising to him, however, was the arrival of a shy researcher from the United States, Karen Strier—just a little lost, but with a determination and friendliness which won his heart and his respect.

I was sure that the best way to protect these monkeys was to study them: to establish a research tradition at Caratinga which would guarantee an understanding of their way of life and their possibilities for survival. Through Russell Mittermeier, this small forest became known worldwide, and through him we had the good fortune to be introduced to Karen. She was evidently the person with the qualities and dedication to begin this tradition, providing a data base on the demography, behavior and ecology of the muriqui which would serve to stimulate long-term studies and provide a field site for other students to obtain experience and develop their research theses, in what was for us a new venture. This brings me to a most important feature of Karen's extraordinary and fascinating study. Besides opening up the muriqui's world for our understanding, she was persistent in her efforts to share her experience and stimulate the interest of Brazilian students in continuing her work. She did not, like so many, obtain her data and disappear; but by working closely with Brazilians, her studies on the details of the muriqui's behavior are now inextricably linked with the protection of these animals and their forest habitat, and most significantly with the development of an expertise and understanding of wildlife studies and conservation amongst numerous Brazilian students. The contribution of her studies extends well beyond the writing of this book. We are grateful to Karen for our knowledge of the fascinating world behind those faces in the forest, for the research tradition she has developed at Caratinga, and for her enthusiasm and dedication in sharing her knowledge with so many people.

CELIO MURILO DE CARVALHO VALLE
Professor of Vertebrate Zoology
Federal University of Minas Gerais
and President of Fundação Biodiversitas

Acknowledgments

Many people have been important to this book, and to the research that led up to it before I ever saw a muriqui. Three of my professors at Swarthmore College—Steven Piker, Allen Schneider, and Tim Williams—influenced the way I thought about animal behavior, and, in doing so, prepared me for the path I followed. It was also during my undergraduate years that I had the privilege to spend six months with the baboons at Amboseli National Park in Kenya. There I discovered for myself the fascination of studying primates in their natural habitat and, thanks to Glenn Hausfater, learned the difference between watching and observing behavior.

I first heard about muriquis from Irven DeVore, my graduate advisor at Harvard University, and he has shared his wisdom, support, and friendship ever since. Also at Harvard, I benefitted from Mark Leighton's guidance, and from his continuing friendship and feedback, as well as that of other colleagues, especially Jim Moore and Barb Smuts.

None of my research in Brazil would have been possible without the help of three people: Russell Mittermeier, who showed me my first muriqui at Fazenda Montes Claros; Celio Valle, who sponsored the first year of my research and championed the long-term project; and Feliciano Miguel Abdalla, who has been a guardian, both to the muriquis and to me during my stays in his forest. By establishing the Biological Station of Caratinga, Sr. Feliciano opened his land to others, who have been allies during difficult times as well as during good ones. These include: Cristina Alves, Priscilla Andrande, Steve Ferrari, Nadir and Lada Ferreira, Jairo Gomes, Rosa Lemos de Sá, Cida Lopes, Sergio Mendes, Ilmar Santos, Andy Young, and Dr. Gustavo Fonseca, who has also sponsored and collaborated in my subsequent research.

Eduardo Veado was the first of the Brazilian students to join me with the muriquis at Fazenda Montes Claros. Francisco (Dida) Mendes, José (Zé) Rímoli, Adriana Odalia Rímoli, Fernanda Neri, and Paulo Coutinho have followed him, participating in the long-term research through their dedicated efforts. Their collaboration and commitment to the muriquis have forged deep links between us, and I am grateful to them for their contributions to the research and to the book. José Rímoli and Adriana Odalia Rímoli donated many of the photographs, and Eduardo Veado provided the drawings. Other dedicated Brazilians who have contributed to the comparative research at the other field sites include: Dr. Bento Vieira de Moura Netto, Sandra Paccagnella, Pedro Luis Rodriguez de Morães, Oswaldo Carvalho Jr., Luiz Paulo de Souza Pinto, and Claudia de Costa. In addition, Sr. José Carlos Reis de Magalhães hosted my visit to Fazenda Barreiro Rico, and both he and another of my sponsors, Dr. César Ades of the Universidade de São Paulo, have contributed to many spirited conversations.

A number of organizations have been generous with their support. The Brazilian research council, CNPq, granted permission for the research; and the National Science Foundation, the Fulbright Foundation, the Joseph Henry Fund of the National Academy of Sciences, Sigma Xi, the World Wildlife Fund, the L.S.B. Leakey Foundation, the Anthropology Department at Harvard University, the National Geographic Society, the Graduate School of the University of Wisconsin-Madison, the Chicago Zoological Society, and the Liz Claiborne and Art Ortenberg Foundation have financed the work.

The original interest of Bill Curtis in this book was carried on by Kirk Jensen at Oxford University Press, and helped make it possible. Karen Bassler created the maps.

Special colleagues in the United States, such as Ken Bennett, Larry Breitborde, J. Mark Kenoyer, Jack Kugelmass, Walter Leutenegger, Diane Lichtenstein, Jon Marks, Kirin Narayan, and Toni Ziegler, as well as my students at Beloit College and the University of Wisconsin-Madison, have been patient during my frantic times before and after recent field trips, and especially during the writing of this book, when it was all I could talk about. In particular, both Margaret Schoeninger, who also commented on an earlier draft, and Chuck Snowdon provided friendship and inspiration throughout.

Frans de Waal was the first to suggest this book soon after he visited Fazenda Montes Claros in 1988, and he generously read and commented on an earlier draft. Shannon Brownlee, who wrote a story about muriquis after watching them with me in 1986, made meticulous editorial and organizational suggestions. Her eye for style and unfailing enthusiasm were essential to this book's completion.

Murray Strier also read and carefully critiqued the manuscript, and his genuine appreciation of every aspect of the research has been an intellectual and emotional buoy over the years. Both he and Arlene Strier encouraged me in this endeavor, just as they have in all others.

Since our days together as students, Rich Summers has been my principal stabilizing influence. In Brazil and in the United States, and in his careful reading of this entire manuscript, he has been a constant source of ideas and new perspectives; in his tolerance of my work and his companionship, he has been like the muriquis described in this book.

Contents

Foreword, vii

Acknowledgments, ix

Introduction, xv

1. Charcoal Monkey, 3

2. Fragmented Forest, 9

3. Models to Mud, 15
Fazenda Montes Claros, 18
First Impressions, 23
Research Design, 27

4. From Days to Years, 35
Close Encounters, 37
Rainy Days, 42
Expanding the Research, 45

5. Early Risers and Other Surprises, 50

6. Peaceful Patrilines, 66

7. Life Histories, Unsolved Mysteries, 82

8. Conservation Concerns and Compromises, 98

Appendix: Forests Known to Support Muriquis, 113

Notes, 115

References, 126

Index, 135

Introduction

This is the story of a vanishing primate, the muriqui monkey of the Brazilian Atlantic forest. It is about what we now know about this highly endangered species, and about what we still need to learn to ensure its survival. Muriquis are extraordinary monkeys, which defy many theoretical predictions about primate behavior. Their striking lack of aggression and the egalitarian relationships between males and females are just two of the traits that set muriquis apart from other primates, offering provocative glimpses into a way of life very different from our own.

This is also a story about conducting field research, told through my own experiences with muriquis spanning the last decade. In 1982 I visited one of the last remaining muriqui strongholds, a 2,000 acre forest at Fazenda Montes Claros in the Brazilian state of Minas Gerais. There I began what has become the only ongoing, long-term study of this species to date.

My principal interests have been to understand the behavior and ecology of muriquis from a comparative perspective, and to collect basic data that will contribute to conservation efforts on their behalf. These two goals have persisted over the years, but a brief incident that occurred early on marked a turning point, when the research became more than a dispassionate study motivated solely by scientific questions.

It was December 16, 1983, six months into the 14 month time period allotted for my doctoral dissertation research. It was a hot summer day and the forest was unusually still. Only the cicadas and the incessant mosquitos seemed immune to the late morning heat. Even the muriquis had stopped feeding on the myrtle berries that had been their primary food source all week. It was not yet noon, but they had already settled down for their midday nap in a cluster of trees near the top of the ridge. I was sitting in the shade of a nearby tree, looking forward to a few hours of calm after the difficult trek that the muriquis had led me on that morning. I could see most of the 23 muriquis in the group from my vantage point, and was systematically recording the spatial relationships between them at 15 minute intervals. The majority of the monkeys had already planted themselves securely along the tops of thick boughs and appeared to be asleep, but occasionally one of them would shift to another position, closer to one of its associates.

A flash of movement caught my eye from the opposite direction, where an unfamiliar male was slowly approaching. As he came nearer, it was clear that he was a male from Jaó, the other muriqui group in the forest. Encounters between the Jaó group and the Matão group, which I was focusing on, were becoming more frequent now that the myrtles along the ridge tops were producing fruit. I had seen this male shadowing the Matão muriquis a few hours earlier, but he had kept far enough away to avoid provoking any reaction from them.

When the Jaó male entered the canopy above me, he suddenly stopped short and began a series of loud, frenzied alarm calls. He had apparently been startled by my presence and began to threaten me, breaking branches and dropping them all around me as he swung wildly about. Four of the resting females from the Matão group immediately rushed over. I knew that muriquis respond to the alarm calls of one another, so the arrival of these familiar females—Nancy, Mona, Didi, and Louise—did not surprise me. It was distressing, however, that they had responded to alarms from a strange male aimed at me because they were already very accustomed to my daily presence. Were the Matão females going to join the Jaó male's threats? I worried that this event was going to cause them to revert to the skittishness that had characterized their original behavior toward me. How long would it take before they began to accept me again?

The females hesitated before they reached the tree with the Jaó male. They huddled together, then looked at the male, then at me, and then back to the male, who never ceased his threats as he solicited the females' support. Seconds later, the females charged toward the male and began to threaten him! The Jaó male froze, as if he, too, had expected a very different reaction. The females lunged toward him, and he fled into an adjacent canopy with the females close behind. They all disappeared down the slope, the male in front, the females behind. It was futile to try to follow them at such speeds, so I stayed put. The forest was filled with the swishing sound of branches as they bent and then rebounded from the muriquis' weight, and the long horse-like neighs and dog-like barks of the females in pursuit. A few minutes later, the females returned to the tree just above me; the Jaó male was nowhere in sight.

The females began to embrace one another, chuckling softly as they hung suspended by their tails, wrapping their long arms and legs around each other. Two of the females disengaged themselves from the others. Still suspended by their tails, they hung side by side holding hands and chuckling. Then they extended their arms toward me, in a gesture that, among muriquis, is a way to offer a reassuring hug.

It took all of my scientific training and willpower to resist the temptation—and the clear invitation—to reach back. I had never touched the muriquis before, and I knew that I could not touch them now and still hope to remain the passive observer that was so essential to my ability to record their behavior for the remainder of the study. Furthermore, human and nonhuman primates can share many of the same diseases and parasites, and physical contact would increase the risk of transmitting something harmful to them.

Soon all four of the females who had come to my defense returned to the rest of the group, where they were greeted by softer neighs as they settled back into their places along the branches. The entire interaction, from the moment the Jaó male approached until the females had returned to their sleeping sites, took less than 10 minutes. But it shaped all subsequent years of the research.

Developing a field study on a poorly known primate is far more complicated than it often appears in televised documentaries, where well-marked trails lead the living-room viewer to a group of well-habituated monkeys engaging in interesting behaviors which the narrator explains in an authoritative voice. The months of planning and sheer physical labor involved in developing a trail system, the long, frustrating days spent trying to locate and follow the monkeys, and the uncomfortable hours in silence, far from family and friends, swatting hordes of mosquitos in hot and humid conditions while the monkeys rest calmly overhead are, in many respects, the most challenging aspects of field research. Yet they are rarely discussed.

In a field study, interesting discoveries very seldom leap out at an observer. Meticulous efforts are required to design methods that record behavior systematically, in ways that will be comparable to other studies and have the power to test specific hypotheses that advance scientific knowledge. Making decisions about what questions to ask, how to ask them, and what data are needed to answer them, as well as the logistical challenges of working in impenetrable areas with difficult terrain, are integral components of all field research.

To understand muriqui ecology and behavior, it was necessary to characterize the muriquis' environment and their adaptations to it. In this book, systematic data on their diet, ranging patterns, and social behavior are integrated with other studies to illustrate how muriquis resemble—and differ from—other primates. Many of the scientific findings described in this book have been reported previously, and references are provided to these sources for those who wish to consult the original results.

Quantitative data are essential for valid comparisons with other studies, and, in the forest, muriqui behavior is recorded according to a carefully developed protocol. But these data alone do not convey what the day-to-day experience of accompanying muriquis has been like, and many special events and interactions elude neat, numerically coded categories. This book includes these anecdotes because it is the stories about the monkeys and the progress of the research that provide an essential context for the scientific findings. I hope that these tales impart something of what following the muriquis has been like over the years, in a way that is accessible to anyone interested in primate behavior without compromising the integrity of the results.

In the beginning, I worked with a single Brazilian assistant, Eduardo Veado. By 1986, however, as international interest in the muriqui continued to increase, the study grew to include what has now become a large team of dedicated students and colleagues. At the same time that the number of researchers has expanded, so has the number of muriquis in the group. The Matão group has nearly doubled in size since the study began. This is a very healthy rate of growth, and an encouraging sign that even a small population of these monkeys can recover if it is well protected. Documenting the changes in the group over the last decade has been fundamental as the research shifts from its original snapshot of a year in the lives of these muriquis to a video that reveals each monkey's individual history.

We have learned more about muriquis in the last 10 years than we had since their scientific discovery in 1806, and there is still a great deal more that they can teach us. International efforts are now dedicated to preserving the muriqui and its unique way of life, but it is sometimes the case that scientific and conservation priorities are at odds, and many difficult choices lay ahead.

The first time I touched a muriqui was through a wire cage. It was January 1989, and I was visiting the Primate Center in Rio de Janeiro. Muriqui fur was much softer than I had imagined it would be after years of observing these monkeys in the wild, and the pads on their fingers were surprisingly tender. I tried to imitate some of the friendly chuckles and reassuring clucks I had heard in the forest, and soon I found myself engaged in a lively exchange.

The three young females I was playing with had come from different areas. One of them had come from a forest close to where I work. I had met her earlier, in July 1988, when she was chained to a post along the side of a house in the nearby city of Caratinga. She had been "found" in April of that year, and her well-intentioned owners were trying to raise her on a diet of bananas and rice. At that time, her eyes were glazed, her bones were visible through matted fur, and her tail was covered with dried waste.

The other two females, from other regions, had similar stories. All had been sent to the Primate Center, where they were receiving superb veterinary care and a well-balanced diet, including fresh leaves collected from an adjacent patch of forest. Eventually these females will become part of a carefully monitored captive breeding program, and perhaps their offspring will ultimately be returned to protected forests—and freedom.

Yet despite my knowledge that the three female muriquis were now safe, and would contribute to the preservation of their species, it was impossible for me not to compare them to the muriquis I have known in the forest. Reflecting on the tragic circumstances that led these females into captivity and that threaten the continued survival of remaining wild populations was what motivated me to write this book.

FACES IN THE FOREST

1

Charcoal Monkey

Everyone admires
The monkey who walks on its feet
The monkey was once human
It can walk as it desires![1]

Muriqui is the Tupi Indian name for the largest South American monkey, the gorillas of the New World, in a sense. Muriquis of both sexes grow up to weigh nearly 15 kilograms,[2] but even their weight fails to convey the impression they give when they are glimpsed in the wild (Figure 1.1). Muriquis measure about five feet when they hang suspended from their long arms, so, although they weigh less, they appear nearly as tall as a human. But unlike the tail-less gorillas and humans, muriquis have long, grasping, "prehensile" tails that are strong enough to support their full body weight while they feed or socialize upside down.

Muriquis are agile monkeys, which is probably why they cannot be any heavier. They travel by swinging hand over hand through the canopy, using their long arms and tails to help propel them along. Muriquis possess only vestiges of a thumb, but their other four fingers are long and slightly curved to hook easily over the tops of branches. Because of these specializations for suspensory locomotion, muriquis can travel much more rapidly than most other arboreal primates, saving them time in their search for high-energy fruits.[3] Indeed, the ability to find widely dispersed fruiting trees in the forest gives muriquis a nutritional advantage over slower primates, whose quadrupedal locomotion precludes them from exploring such large areas for food.

Muriquis may be forced to feed on leaves when fruits are seasonally scarce. To cope with this problem, muriquis possess a number of physical traits, including small incisors, large, shearing molars, and robust jaws, that are characteristic of other primates known to be adapted to a diet filled with leaves.[4] Sharp, crested molars help to break open the leaf's cell walls, freeing the nutrients that are stored there, while thick jaws and jaw muscles are important for chewing. Muriquis also have pot bellies, which give both males and females the appearance of always being pregnant,

Figure 1.1 Helena and Tereza—both adolescent females—in a friendly hug (photo by A. O. Rímoli).

but are actually due to the disproportionately large size of their intestinal tract, a trait that is important for absorbing nutrients from leaves (Figure 1.2).[5] Even the muriquis' large body size, which is associated with a relatively lower metabolic rate, may be linked to a heavy reliance on leaves during at least part of the year.[6]

Male and female muriquis resemble one another in more ways than their bellies. There are nearly no sex differences in body size or canine size, and if it were not for their exceptionally large, visible genitalia, it would be difficult to tell males and females apart.[7] Most other primates which live in the same type of multimale, multifemale groups as muriquis exhibit a high degree of sexual dimorphism, with males generally larger than females in both body and canines.[8] Perhaps because they have size on their side, males in these species are also generally able to dominate females. In muriquis, however, females have nearly the same size advantages as males, and cannot be physically harassed or threatened.[9] The muriqui is an example of an unusual species in which females are very much in control.

To many Brazilians, the muriqui is known as *o mono carvoeiro*, or the charcoal monkey, because its golden-gray coat and sooty face resemble *carvoeiros*, people who burn wood for charcoal.[10] Ironically, the clearing of the Atlantic forest for charcoal, timber, and agricultural land has reduced the muriqui's habitat to just a fraction of its original extent, and

Figure 1.2 The muriquis' large gastrointestinal tract enables them to absorb nutrients from leaves.

today fewer than 500 muriquis are known to persist in less than a dozen isolated forest fragments.[11] The perilous condition of these remaining muriqui populations has attracted considerable international attention, and was a strong impetus for beginning this study.

Scientifically, the muriqui is called *Brachyteles arachnoides*. Muriquis are members of a large family of New World monkeys, the Atelinae, that includes woolly monkeys (*Lagothrix*), spider monkeys (*Ateles*), and howler monkeys (*Alouatta*).[12] Muriquis are most closely related to the woolly monkeys and the spider monkeys, and, for a time, muriquis were known as "woolly spider monkeys" to reflect this relationship as well as their general appearance (Figure 1.3).

Of the Atelinae, the cat-like howler monkeys have the widest distribution. At least one of the six species of howler monkeys can be found in any sizeable forest from Central America to Northern Argentina, and throughout the Amazon. The brown howler monkey (*Alouatta fusca*) shares the Brazilian Atlantic forest with muriquis. The four species of spider monkey have the next largest distribution, overlapping the central

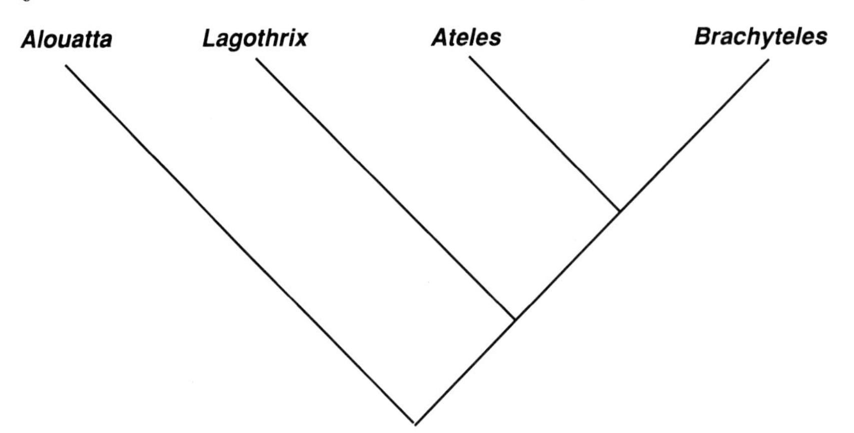

Figure 1.3 Phylogeny of the Atelinae family.

and northern parts of the howler monkey's range but not extending south of the Amazon. The two woolly monkey species can also be found in pockets of the Amazon forest, but the muriqui, the sole member of the genus *Brachyteles*, has the narrowest distribution of all.[13]

The muriqui and its family, along with all other monkeys, apes, and humans, are members of the Haplorhine primates.[14] The monkeys are split into two groups: the Old World monkeys, which are found throughout the tropical forests and savannas of Africa and Asia; and the New World monkeys, which are found only in the forests of Central and South America. The earliest monkey fossils, which are 30–35 million years old, come from the Fayum region of Egypt, and some researchers believe that both New World and Old World monkeys descended from this common monkey stock.[15] But whatever their origins, the New World monkeys separated from their African relatives long before the first human ancestors, or hominids, appeared.

Spurred by their similar physical features and antics, humans have long been attracted to other primates, and we rightly consider them our cousins on this planet. In Brazilian folklore, monkeys are believed to have once been mischievous humans who were punished for their pesky deeds, a view that turns modern evolutionary theory on its head.[16] But all primates, including humans, have changed over the millennia in response to environmental pressures and random genetic factors, and few scientists today look at nonhuman primates as living models of early human ancestors. Nevertheless, understanding the behavior of monkeys and apes provides insights into how long-lived, socially complex animals like ourselves are influenced by their environment. Understanding how and why nonhuman primates differ from one another should provide a deeper appreciation of our place in nature.

Many early primate field studies focused on Old World monkeys, such as baboons and macaques, because these monkeys have adapted to the ecological pressures of life on the ground, just as our hominid ancestors had at least four million years ago. The African apes—chimpanzees and gorillas—also received a great deal of attention, in part because of their astonishingly close genetic relationship to ourselves.[17] But during the last two decades, field research has expanded to include an increasingly broad array of species, such as forest-dwelling Old World monkeys, the New World monkeys, and prosimians. Together, these primates present a much more diverse set of behavioral and ecological adaptations, and, with the comparative studies that are now possible, we can begin to look for patterns that help to explain both individual species' differences and common trends that unite them.

The development of research on nonhuman primates has parallels in most fields of science. In the beginning, researchers focused on descriptive, ethnographic accounts of the animals, which were interesting to read but difficult to compare with one another because nonstandardized methods of observation were used. With increased understanding, the methods of gathering information have become more refined and systematic, and the analyses have become more quantitative. Interpretations of early findings have now been expanded and revised as long-term studies on familiar species yield more rigorous data, and studies of new species in different habitats provide a comparative backdrop against which general theories can be formulated and tested.

Contemporary primatologists bring a much broader perspective to their studies than was possible even two decades ago, and carry a keen awareness that quantitative methods are necessary if one wants to make valid comparisons. As a result, primate studies can now entertain multiple levels of comparisons, ranging from those between individuals in a study population, to those between populations of the same species in different habitats, to those between species.[18]

The roots of primatology are interdisciplinary, growing out of experimental psychology in the United States, gestalt psychology and zoology in Europe, and anthropology in Japan. Evolutionary biology and ecology are also prominent in primatology, but this is a relatively recent development. Today, many studies attempt to distinguish the inertia of phylogeny against the breakthroughs of ecological adaptations. Presumably, ancestral traits will be common to most related species, while traits that are uncommon, occurring only in some species in specific environments, are candidates for more recent evolutionary events.

Untangling the muriquis' unique suite of traits will provide insights into the evolutionary and ecological factors that have shaped both muriquis and, by comparison, other primates' development. But there is an additional urgency underlying primate field research today that is as stimulating and motivating as the scientific challenges. Human population growth and its associated intrusion into the world's tropical forest habitats have threatened the continued existence of our study subjects. Of some 233 living primate species now recognized, nearly 50% are listed in the International Union for the Conservation of Nature (IUCN) Red Data book because they are in some danger. Twenty-five percent are classified as endangered, meaning that their numbers have dropped to such critically low levels that they will become extinct unless drastic measures are taken to protect them.[19] The muriqui is one of the most severely endangered primates, and understanding its behavior and ecology in the wild is inextricably linked both to its past, and to its future.

2

Fragmented Forest

Scattered throughout southeastern Brazil, from Bahia in the north to São Paulo in the south, and inland into Minas Gerais, are isolated stands of what remains of the Brazilian Atlantic forest. It once carpeted the chain of mountains running along the eastern coast of Brazil from the states of Rio Grande do Sul to Rio Grande do Norte, an area of roughly one million square kilometers. Agriculture and logging have reduced this unique ecosystem to less than 5% of its original extent.[1] Now all that is left are fragments of a once magnificent forest that early explorers described as a living paradise (Figure 2.1).

The Atlantic forest is not merely an extension of the Amazonian rainforest. It is physically distinct, with its own exceptional evolutionary history and communities of flora and fauna.[2] In the beginning of the Quarternary, at 1.64 million years ago, climatic changes led to periodic shrinkage and expansion of the world's great rainforests. Pollen and sediments from the Amazon indicate that large areas of modern-day rainforest were open grassland at various times during this period. Rocks and sediments found in the Atlantic forest's Serra do Mar, in São Paulo state, also indicate cyclical periods of semi-arid and humid climates in these "mountains of the sea" that correspond to the glacial and interglacial epochs in the northern hemisphere.

As climate fluctuated, the contiguity of the Atlantic forest was disrupted, creating isolated forest islands known as refuges.[3] The lack of genetic exchange between species in separate refuges permitted the evolution and differentiation of numerous species of birds,[4] reptiles,[5] plants,[6] and primates,[7] resulting in high levels of endemism, or organisms found exclusively in one area.

During dry periods, when the contiguous Atlantic forest contracted into central refuges, the animal populations were cut off from one another for variable periods of time. Some populations accompanied the re-expansion of the forest during wet periods, and intermingled with other members of their species, while others had differentiated during isolation to the point that, even when the geographical barriers were removed, they no longer mated with former members of their species, and thus evolved

Figure 2.1 Most of the remaining forests in the state of Minas Gerais occur on hilltops, surrounded by agricultural and pasture land (photo by K.B. Strier).

into distinct species or subspecies.[8] Still others never recovered their original distribution because of competition or poor migratory abilities, and remained in the remnants of their refuges. Identifying the precise locations and limits of these ancestral refuges is a matter of comparing fossil records against current distributions of species. Areas with the greatest numbers of both fossil and current species indicate central areas, or refuges, where species were forced to aggregate during periods of forest contraction.

The distributions of the six primate species endemic to the Atlantic forest suggest three central refuges: the Bahia refuge, bordered by the Rio das Contas and Rio Pardo in the state Bahia; the Rio Doce refuge, north of the Rio Doce in what is now the northern part of Espírito Santo; and the Paulista refuge, which encompassed the upper drainage area of the Rio Tietê in the state of São Paulo.[9] Historical records show that muriquis have always inhabited both the Bahia and Paulista centers, but there is no evidence that they have even occurred in the Rio Doce center (Figure 2.2).

The Bahia and Paulista refuges lie at opposite ends of the southeastern Atlantic forest, and it seems that muriquis spreading west and north from the Paulista refuge and south from the Bahia refuge never re-established contact once climatic changes caused the Atlantic forest to contract. In fact, some taxonomists have suggested that there are two races or subspecies of muriquis, *Brachyteles arachnoides arachnoides* and *Brachyteles arachnoides hypoxanthus*, which correspond to the two different

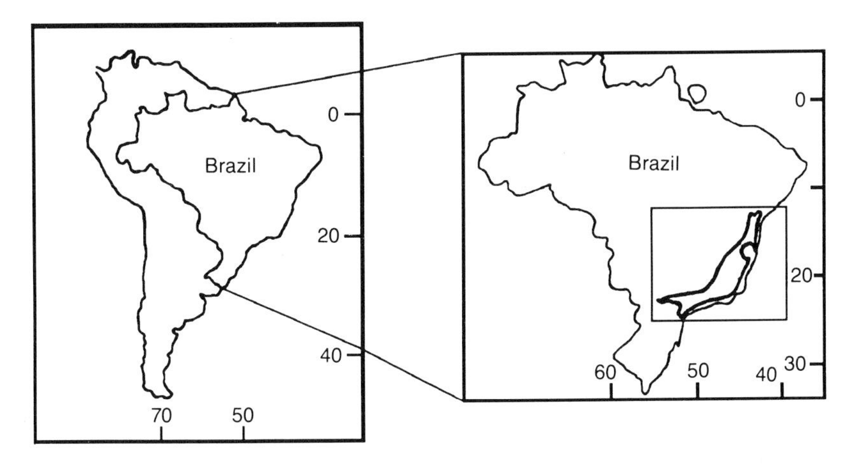

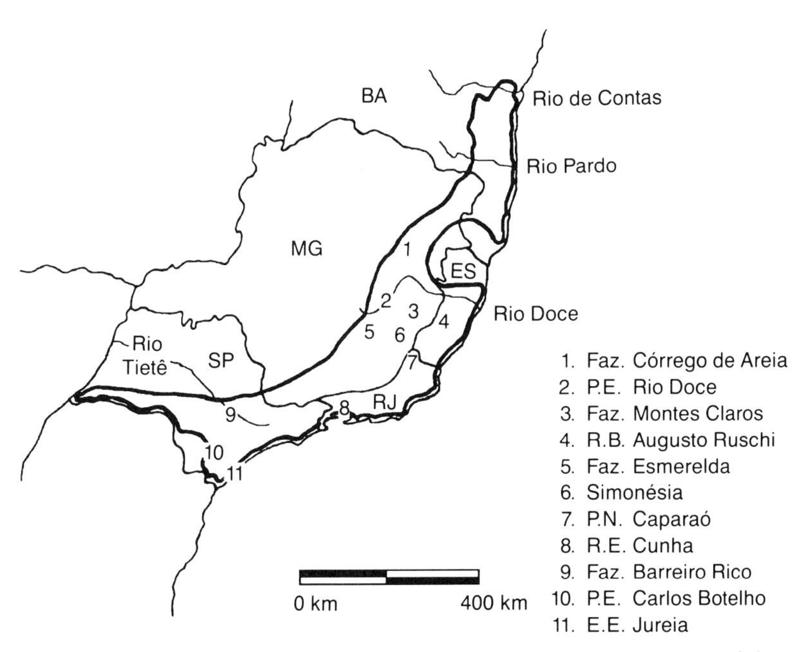

Figure 2.2 Original extent of the Brazilian Atlantic forest, and the locations of the 11 forests that currently support muriquis (see also Appendix).

populations.[10] There is some morphological evidence to support this view, including the fact that muriquis in São Paulo from the Paulista refuge have dark faces, while those in Minas Gerais and southern Espírito Santo from the Bahia refuge have individually distinct pink and white mottling mixed with black on their faces. Most researchers, however, believe that insufficient generations have elapsed for the species to differentiate so significantly.

Human beings, too, have had an impact on the historical distribution of the muriqui. The Portuguese exploration of southeastern Brazil began in the 1500s, but ports and trading posts were established mostly on the coasts of Bahia and Rio de Janeiro, and the only humans to occupy the southeastern Atlantic forest inland were the diverse Botoccudo Indians.[11] By the 16th and 17th centuries, sugar cane plantations on the fertile coastal plains supplied nearly all of Europe's sugar. European settlers struck inland with the discovery of gold and diamonds in the late 16th century, and by the 1700s mining towns were beginning to be transformed into agricultural settlements. All of this development was, of course, eating into the muriqui's habitat. Yet, in the 1800s, there were still extensive forests inhabited by muriquis.

During his travels through the region in 1815–1817, the German explorer, Prince Maximilian Wied-Neuwied, described crossing immense forests, "... full of gigantic trees excellent for all kinds of construction...."[12] Wied-Neuwied also made numerous references to muriquis in his writings. "The giant of the monkeys,"[13] as they were called, provided an abundant source of meat during much of his expedition. Wied-Neuwied recalled observing a large group of muriquis crossing from branch to branch overhead, before killing three individuals.[14] The Prince's party was not the only one to take advantage of the muriquis, however, since Puris Indians were often encountered ornamented with muriqui skin,[15] and roaming European hunters came to use its skin as protection from the rain and as purses for storing bullets.[16]

A short while after Wied-Neuwied's exploration had ended, continued pressure from European settlers resulted in the annihilation of the indigenous peoples of the Rio Doce Valley, and iron extraction in this region began. In 1861 the first railroad was constructed through the Atlantic forest in the state of Minas Gerais to transport the increasing production of coffee and sugar. In 1910 the first steel mill was established in the city of Ipatinga, also in Minas Gerais.

The final phase of destruction of the Atlantic forest occurred in the beginning of the 20th century, when enormous tracts of forest were cleared for crops and pasture and southeastern Brazil became the agricultural and industrial center of the country. Originally, over 80% of the state of São Paulo was forested, but by 1979 only 7% of the forest remained. In the 1930s, over 50% of state land in Minas Gerais was forested. By 1980 less than 5% of this forest remained, and today much of what is left of the forest in this state occurs on private farms and ranches, mostly on hilltops that are of marginal agricultural value. The region previously covered by the Atlantic forest in southeastern Brazil represents only 11% of the country's territory, but over 40% of the Brazilian population now lives in this area.[17]

The accelerating destruction of the Atlantic forest has had a devastating effect on the muriqui. Wied-Neuwied's writings provide convincing evidence that muriquis were once extremely abundant. He concluded that muriquis "prefer forests from the high plain to the dry mountains covered with low forest,"[18] which, when extrapolated back to the historical size of the Atlantic forest, yields an original population estimate of about 400,000 individuals.[19] But by the time Brazilian scientist Alvaro Coutinho Aguirre conducted a thorough survey of remaining muriqui populations in the 1960s, the number had dropped to fewer than 3,000 animals. In a subsequent census in 1972, the population had dwindled to only 2,000 individuals as more fragments of forest, particularly in southern Bahia, disappeared.[20] In 1987, the total known population of muriquis was 386, scattered in just 11 locations (see Figure 2.2).[21]

Over 50% of the entire remaining Atlantic forest is confined to the coastal mountains of São Paulo, the state that supports more muriquis than any other. The steep terrain has made census taking difficult in these mountains, and, while this has probably resulted in an underestimate of existing muriqui populations, it has probably also contributed to their continued survival.[22] Muriquis also occur in the states of Espírito Santo and Minas Gerais, but they seem to have been lost from the states of Bahia and possibly Rio de Janeiro.

The forest fragments in which muriquis survive today are extremely variable in size, degree of human disturbance, and protection status (see Appendix). They range from the tiny, 100 acre forest located on Fazenda Esmeralda (a private ranch in Minas Gerais), to the impressive 94,000 acre Carlos Botelho State Park in São Paulo. Some of these forests have experienced natural disturbances such as lightening fires; others, especially the forests belonging to individual landholders, are subject to varying degrees of selective logging and hunting activities. These activities may have serious consequences for the muriqui inhabitants: the muriqui population at Fazenda Esmeralda has declined 33% from a documented high of 18 individuals,[23] while efforts to locate muriquis at Fazenda Corrégo de Areia in 1990 were unsuccessful, and it is not certain whether muriquis still survive at this site.[24]

The prospects for some of the other remaining muriqui populations are much more encouraging. The forest at Simonésia, for example, was purchased from its owner in 1991 by the Brazilian conservation foundation, Biodiversitas, which will now be able to protect the muriquis there. The forest at Fazenda Montes Claros, where most of my muriqui research has focused, still belongs to the owner of the farm, Sr. Feliciano Miguel Abdalla. In 1983, Sr. Feliciano donated a 2.5 acre piece of land with a house to the Brazilian Foundation for the Conservation of Nature.[25] This has become the Biological Station of Caratinga; it stands at the entrance to the central part of the forest, symbolizing Sr. Feliciano's commitment to the protection of the forest and the research it supports.

The muriqui population at Montes Claros has nearly doubled in size during the decade spanned by my research, suggesting that improved habitat protection can lead to an increase in muriqui numbers throughout its remaining distribution.[26] In fact, the comparative data that are beginning to accumulate indicate that mildly disturbed forests may actually be able to support higher muriqui population densities than more pristine forests can.[27] Continuing studies must still confirm these preliminary findings, but, at the moment, the future of the muriqui may be brighter than we once thought.

3

Models to Mud

In the early 1980s, primatologists began to get an intellectual grasp on how ecology has shaped primate behavior. Two seminal papers by Richard Wrangham, a well-known primatologist whose major research interest is chimpanzees, pulled the comparative data that had accumulated for primates together with evolutionary and ecological theories.[1] Many of the principles had been identified before for ungulates, birds, and bats,[2] but Wrangham was the first to apply them to primates in any cohesive way, developing a model that served as a theoretical blueprint for an entire generation of field research.

Originally, Wrangham considered the resources that most directly influence male and female reproductive success, or their genetic contribution to future generations. These important resources differ for males and females because of fundamental biological differences between their reproductive potential. During gestation and lactation, a female does not reproduce, so the number of offspring a female can produce in her lifetime is ultimately limited. But, within these limits, a female's reproductive success is strongly related to the quality of her diet. There is convincing evidence from a number of captive and field studies that better-fed females begin reproducing at an earlier age, produce healthier offspring, have shorter interbirth intervals, and live longer to reproduce.[3] Male reproduction, by contrast, is limited primarily by the number of females a male can fertilize. Thus, while the most important resource for females is access to high-quality foods that will improve their nutritional condition, the most important resource for males is access to sexually receptive females. According to this model, the ways females exploit their important food resources should determine female behavioral strategies, which, in turn, should determine male behavioral strategies. The complex nature of this hierarchy between female and male strategies is reflected in the number of different social organizations exhibited by primates today.

Wrangham then went on to detail the ecological variables that lead to the formation of what he termed "female-bonded" groups, where females live with their close female relatives and cooperate with one another in defending important food resources from other groups of related females. Wrangham argued that females would remain in cohesive groups with their female relatives only when preferred "growth" foods, such as energy-rich fruits, were sufficiently abundant that females could feed together without divisive competition. Under these circumstances, female-bonded groups would be beneficial because larger, cooperative groups of females would have a competitive edge in defending important fruit resources. But, when these preferred fruit sources become seasonally scarce, females would have to be able to shift their diets to include more abundant, evenly distributed "subsistence" foods such as grasses and leaves. Cooperative defense of any remaining large fruit patches would still make it beneficial to females to stay together during these times. Males would distribute themselves in response to the distribution of these female-bonded groups. If female food resources were sufficiently clumped, a single male could help defend a small territory that contained the food that the females needed. But if food, and, consequently, females, were distributed over too large an area for a single male to monopolize, several males might compete amongst themselves for mates whilst accompanying female groups in their search for food.

At the time that this model was being developed, the majority of long-term data on primates had come from studies of Old World monkeys, which were almost all female-bonded. A few monkey species, and all of the apes, were exceptions to the female-bonded system, but these could be explained by explicit deviations from the ecological correlates outlined in the model. Chimpanzees, for example, have fluid grouping patterns because their preferred fruit resources occur in variable-sized patches. Consequently, female chimpanzees fission into small or solitary foraging units when the fruit patches are small, and aggregate only when there is sufficient food to fill more mouths. When female chimps disperse into their widely separated core areas, individual males could roam between them, but then they would not be able to defend the females against invading males from other communities. A better strategy for males is to stay in their natal groups and cooperate with their brothers to keep other, unrelated males out of their community range and away from their females. This way, even if a male is unable to reproduce himself, at least some of the genes which he shares with his relatives will be passed on indirectly.

Gorillas, by contrast, live in unimale or multimale, multifemale groups that travel together as cohesive units. Their primary food resources are abundant shoots and stalks, and there are few advantages to females to remain in their natal groups because there is not much need for cooperative resource defense. But, because there is comparatively little competition between females over these plentiful foods, there are few disadvantages of associating together. Consequently, one or more males can herd an aggregation of females, defending them from other, marauding males looking for mates. Indeed, female gorillas living in groups protected by dominant silverback males, may actually increase their feeding time because the males in their group prevent other males from harassing them.

Wrangham's model, which related female behavior to the distribution of food resources and male behavior to the distribution of females, can be used to generate a number of testable hypotheses about primate grouping patterns and social organization. All that is necessary is to determine what a primate prefers to eat, whether it can shift its diet when preferred foods are scarce, and how both preferred and subsistence foods are distributed. Females should behave in ways that are consistent with the ecological parameters outlined in the model, and males should respond to the behavior of females. If food resources fit the pattern for female-bonded species, then the social organization should be characterized by affiliative relationships among females that remain in cohesive kin groups and competitive relationships among unrelated males. If the distribution of food resources diverges from the female-bonded patterns, then chimpanzee- or gorilla-like systems should occur, depending on the degree of female feeding competition.

It was apparent that examining these variables in a poorly known species would be an ideal way to test the predictions of this model, and, when I heard about muriquis during my second year of graduate school, I knew I had found my study subjects. At that time, international attention had begun to focus on the conservation status of the muriqui, but no one had yet described their behavior or ecology in any detail. Studying muriquis had an additional appeal, because the same data on feeding, ranging, and social behavior that were necessary to evaluate the theoretical predictions were also important to the development of informed management plans on the muriquis' behalf.

I was concerned, however, that it might not be possible to obtain these critical data about muriquis. The very reasons that made muriquis such interesting study subjects also made the task of studying them seem nearly hopeless. Before it would be possible to collect basic data on muriqui diet, ranging behavior, grouping patterns, and social relationships in a way that would permit comparisons between them and other primates, it would first be necessary to habituate a study group, identify individuals and accumulate data on their kinship, and define their behavioral repertoire. I knew that initiating a new study on a new species would be complicated, and that a better idea of both the animals and the field conditions was needed before any study could be planned.

I contacted Russell Mittermeier,[4] then director of primate research at the World Wildlife Fund, and arranged to accompany him on his upcoming trip to Brazil in June 1982. Mittermeier was taking a few other American scientists representing different conservation agencies to Fazenda Montes Claros, the forest where he felt that the kind of study I proposed was most likely to succeed. In addition to seeing the muriquis there and evaluating the feasibility of the research, Mittermeier would also introduce me to some of the prominent Brazilian scientists and conservationists who had worked in the area: Professor Celio Valle, of the Federal University of Minas Gerais, Dr. Adelmar Coimbra-Filho, of the Rio de Janeiro Primate Center, and Admiral Ibsen de Gusmão Cámara, of the Brazilian Foundation for the Conservation of Nature. These three men had been instrumental in attracting attention to the plight of the muriqui and the Atlantic forest, and their collaboration would be critical to the success of any research effort.[5] The experience of Celio Valle and his students, who observed muriquis intermittently at Fazenda Montes Claros and other forests,[6] would be particularly important to the development and implementation of my study.

FAZENDA MONTES CLAROS

Fazenda Montes Claros is a coffee plantation located along the dirt road that runs between the city of Caratinga and the smaller town of Ipanema in the state of Minas Gerais. Caratinga has grown up around the crossroads between the north–south highway that connects Rio de Janeiro with Salvador, and the east–west highway that runs between Belo Horizonte and Vitória. Driving south on the road from Caratinga to Fazenda Montes Claros, one passes numerous small farms with coffee plants stacked neatly up hillsides that once supported continuous forest, and ranches where cattle graze on newly cleared land. The heavy summer rains each year erode deep crevices into the barren landscape, and the dirt road becomes impassable for days at a time. But in the dry season, when I made my first trip to the forest, it was difficult to see through the red dust that hung suspended in the air.

The 2,000 acre patch of forest at Montes Claros is also surrounded by coffee plantations and pasture, but the forest has been protected since 1944 when its owner, Sr. Feliciano Miguel Abdalla, established himself in the region (Figure 3.1). Everyone thought he was crazy at the time, but Sr. Feliciano loved the forest and has resisted all pressure to clear it for additional crop and pasture land. Despite the enormous profit he could earn from the valuable lumber, he has persisted in removing only those trees that he needs for construction. Inadvertently, he has preserved a natural treasure for science.

During my first visit, I stayed at Sr. Feliciano's house, situated about

Figure 3.1 Sr. Feliciano Miguel Abdalla, standing in his coffee fields in 1982. The western edge of his forest is visible on the slope behind him (photo by K.B. Strier).

2 kilometers from the main part of the forest. On climbing the narrow dirt road from his house to the forest, the air grows noticeably more humid. In fact, Sr. Feliciano confided that one of the reasons he had preserved the forest was because the humidity it traps helps his coffee plants thrive.

The road from the farm continues through one of the central valleys in the forest, until it emerges into more coffee fields and pastures on the opposite side. Except when it passes through the forest, the road is dotted with the houses of the 20 families employed to tend Sr. Feliciano's fields.

The forest at Fazenda Montes Claros comprises a series of hills and valleys ranging in altitude from 400 to 640 meters above sea level (Figure 3.2). It is a mosaic of vegetation resulting from past disturbances and natural causes.[7] Tropical trees from the bean family (Leguminosae) dominate the forest, followed by laurels (Lauraceae), cashews and sumacs (Anacardiaceae), catalpa and bignonia (Bignoniaceae), and figs (Moraceae). Undisturbed primary vegetation comprises roughly 16% of the forest, and is found mainly in the valleys where large hardwood trees, worth over $10,000 each for their lumber, reach more than 30 meters into the sky (Figure 3.3). Various stages of secondary and regenerating forest with smaller trees make up 55% of the forest, and occur along the slopes where selective logging has disturbed the habitat or the sandy soil cannot support larger trees. The rest of the forest is scrub and bamboo, particularly along the dry ridge tops, where evidence of natural fires from

Figure 3.2 The forest at Fazenda Montes Claros (photo by K.B. Strier).

the past remains. The tall canopy in the valleys and on the lower slopes prevents light from penetrating, and in places that have never been disturbed the ground is open and easy to move through. Along the upper slopes, ridge tops, and in places where the canopy has been disrupted, the understory is full of tangled vines and thorny bushes. Moving through these areas is very difficult for humans, but the primates have clearly adapted.

The forest supports three primate species in addition to muriquis: the small, 300 gram buffy-headed marmoset (*Callithrix flaviceps*); the intermediate-sized, 3 kilogram black capuchin monkey (*Cebus apella nigritus*); and the somewhat larger 4.5 kilogram brown howler monkey (*Alouatta fusca*).[8] Both the marmoset and the howler monkey, like the 12–15 kilogram muriqui, are endemic to the Atlantic forest, and, like the muriqui, they are highly endangered.

Two groups of muriquis inhabit the forest at Fazenda Montes Claros. They are named after the major valleys that they are known to inhabit. The Matão group frequently crosses the dirt road that bisects the Matão valley, using the forested slopes on both sides as well as a continuing stretch of forest to the south. The Jaó group uses the Jaó valley and slopes to the north. I later discovered, however, that the Jaó group also uses a large portion of the Matão group's home range, and that Jaó males have begun to make increasingly frequent incursions into all parts of the Matão group's area. The Matão group was the larger of the two when the study began, consisting of 22 individuals including six adult males, eight adult

Figure 3.3 Large trees in the valleys block out sunlight, leaving the understory open and easy to walk through (photo by A.O. Rímoli).

females, two juvenile males, and six infants. I was not able to count the Jaó group until my return in 1983, when it consisted of 18 individuals including 12 adult males, four adult females, and two immatures.

In 1982 there were already several trails in the forest running up on both sides of the Matão valley road. These were created when large trees deep in the forest were removed. Usually a single tree was cut with a hand or gasoline-powered saw, and then removed by a team of oxen, but in the process of such selective logging, the surrounding vegetation is often trampled or cleared. These old logging trails, which originate from the road, provide easy access into the forest, where the muriquis spend most of their time.

The Matão muriquis crossed the road at least once each week in areas where the canopy was still continuous from one side to the other. They were accustomed to seeing the local farmers passing along this road, and for this reason I selected them as the focus for my study. Usually, the muriquis moved up the slopes after crossing the valley, and away from the road they were much more wary because humans rarely followed them there. Sometimes it was possible to maintain contact with the muriquis from one of the old logging trails, but usually they moved away from these trails, deep into forest which, at that time, seemed impenetrable.

During his brief stay at Montes Claros in 1977, Akisato Nishimura marked and mapped the existing logging trails.[9] By 1982, however, many of the trails on Nishimura's map were overgrown, while new trails had been opened for removing additional trees. An updated map would be essential for any future research, and constructing one was an important goal of this first field trip. I enlisted the help of Ilmar Santos, one of Celio Valle's students from the Federal University of Minas Gerais, to measure and re-mark all of the existing trails that passed through parts of the Matão group's home range. I spoke very little Portuguese then, but fortunately Ilmar spoke some English. He tried to teach me Portuguese by singing translated versions of old Beatles' songs during our tedious work, which led to some rather unusual conversations.

In addition to measuring the distance from point to point, we used a compass to determine the curves in the trails and a clinometer to calculate altitudinal changes. We carried a compass, clinometer, and 50 meter tape, but the most cumbersome object was a 1.5 meter sighting post made of heavy wood. One of us would walk ahead carrying the post and one end of the measuring tape until we had gone as far as we could without being lost to view to the other, who stayed behind holding the other end of the tape. Then we would stop, and the person behind would note the compass direction and the slope of the hill while we stretched the tape measure tightly between our fixed positions. Before moving on we would mark our spots with brightly colored plastic surveyor's tape. Each trail was identified by a different letter, and each point was numbered consecutively.

It took half a day to measure and mark an average 800 meter trail, in part because the twists and turns and steep inclinations in the trails meant that some of our sightings were less than 10 meters apart. Some trails ran straight up a ridge where they intersected with other trails; others wound up a ridge and down the opposite side, or stopped at the top.

The 14 days we spent measuring the trails helped me to understand the multiple ridges and valleys that defined the part of the forest that the muriquis used most often. This time was also important in helping me to identify places where new trails were needed. Occasionally we encountered muriquis during our work, and I noted which direction they moved off in and where a connecting trail would be most useful.

FIRST IMPRESSIONS

My first days in the forest were full of unexpected sensations. Along the Matão valley road, trees rose tall, cathedral-like, on both sides. Sunlight splashed against leaves and left bright puddles on the ground. In some places the canopy closed completely overhead, providing an arboreal bridge across the road. It was surprisingly noisy: parrots and woodpeckers, a frog or toad now and then, and always the humming and buzzing of insects.

A bird scurrying through the dry leaves just beside the road upset a thin branch, which lunged out like a snake. My heart froze and I leapt to the other side of the road before realizing my ridiculous mistake. The idea of snakes was still terrifying, and I saw that my low-cut work boots would not be much protection. I pulled out my shiny vinyl-covered pocket notebook, stretched the tight binding open to the back, and wrote BOOTS on the top of the page that had just become my shopping list.

A branch sagged overhead and a low throaty growl began to emerge from the reddish body sitting on it. A male howler monkey had issued this call, and behind him, sitting on smaller branches in another tree, were the smaller blacker forms of the females and juveniles in his troop. I pulled out my binoculars, and tried to focus them on my first primates in the Atlantic forest. It took a while, and by the time the binoculars were adjusted to my eye width and in focus, the howler monkeys had moved out of view.

Mittermeier and the other Americans had gone on ahead, and suddenly I was alone.[10] I walked on slowly, concentrating on the sounds and the sensations of this new world. Chatters and coos, and then lots of small brown monkeys darted into a tree at the edge of the road, and then quickly scampered back into the forest and out of view. These were the capuchin monkeys, or *macaco pregos*,[11] which had littered the ground with half-eaten seed pods.

I caught up to Mittermeier's group, who had heard the commotion and stopped to wait for me. We continued along on the road; I had only a vague idea of what we were really looking for, but I looked anyway. My neck was beginning to ache from stretching it upward in this unfamiliar position. I was glad that I wasn't burdened with the heavy camera bag that the photographer, Andy Young, had slung on his shoulder. Seeing Mittermeier walking along in his running shoes and shorts made me feel a bit foolish for my earlier thoughts about thigh-high boots.

After about 1.5 kilometers, we came to the end of the forested portion of the road, and found ourselves staring out at coffee bushes on one side and a pasture on the other. The forest extends up the hillsides on both

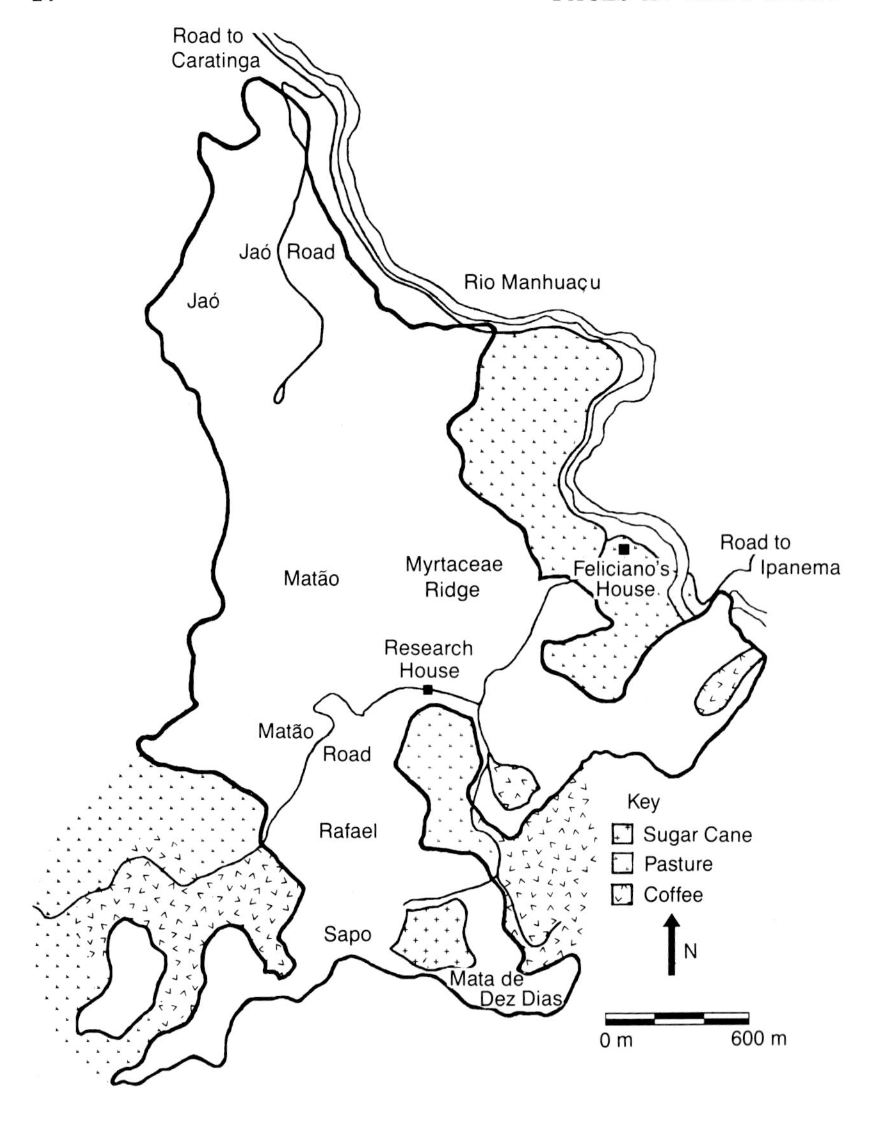

Figure 3.4 The forest at Fazenda Montes Claros, with the location of the main areas in the Matão group's home range.

sides of the Matão valley, and runs along the spines of the ridges where it connects with the forest of another farmer and then abruptly disappears (Figure 3.4). Behind the southern ridge is a part of the forest named Raphael, and, behind it, Sapo. Behind the northern ridge is the Jaó part of the forest, an area nearly the size of the Matão, Raphael, and Sapo forests combined.

After lunching with Sr. Feliciano, we returned to the forest and entered along an old logging trail that began about halfway along the Matão road. A huge tree called a *Jequitibá*, of the famous Brazil nut family (Lecythidaceae) stands like a 40 meter sentinel at the base of this trail. Almost immediately, the trail begins to climb so steeply that in some places it was necessary to pull ourselves up with our hands. The lower part of the trail is lushly covered with red-flowering bushes. A loud buzzing nearby made me halt, afraid of the bees I was allergic to, only to see a humming bird whose wings moved so quickly they were nothing more than a blur. I was beginning to realize that most of my fears in the forest were unfounded.

Climbing in a single file up the trail, our group made too much noise to hear anything more than the dry leaves crunching underfoot. The vegetation was too dense to see through, and it was difficult to stop along the sharply angled slope. How would we find the muriquis if we couldn't hear or see beyond ourselves? We paused periodically, presumably to listen, but we were all out of synchrony and it was never really quiet. I tried to follow Mittermeier, stopping whenever he stopped even if it meant he was standing comfortably on a flat spot while my toes dug into the slope for support. All of these things—how to walk through the forest with minimal noise, how to hear the sounds I didn't yet know—were unfamiliar skills that none of my careful reading or prior years of study had prepared me for.

When we finally reached a clearing midway along the trail, we sat down to listen and rest. The mid-afternoon forest was suddenly still. The forest was as surprisingly silent as the roadside had been noisy that morning. Was this because it was later in the day now, and all of the animals were resting, or because sounds travel differently and are absorbed in the heart of the forest?

We agreed we would have a better chance at finding the muriquis if we split up into two groups. Mittermeier and his assistant, Carlos Alberto,[12] discussed it among themselves in rapid Portuguese that I couldn't yet follow, but I wasn't surprised when they announced that Mittermeier and I would climb to the top of the trail, while Carlos Alberto stayed behind with the others. Mittermeier set off ahead, while I tried to follow just far enough back so he wouldn't hear me huffing and puffing. After what seemed like an hour but was actually less than 10 minutes, the trail began to level off. Soon we reached the top of the ridge, the highest point in the forest, nearly 700 meters above sea level. Turning to survey the forest below, the complex sea of green ridges and valleys seemed to roll into themselves like waves.

The opportunity to see the forest from this vantage point more than compensated for the near-vertical climb. It was obvious now why this forest was so important: off in the distance on all sides the forest stopped abruptly, replaced by a barren, eroded landscape. The forest was an island, rich in unknown, unstudied, unimaginably beautiful life. For the first time, the desire to really know this forest began to take hold in my mind: to live here, to walk through it daily, to become a part of it. It was a remarkable privilege to be here, but where were the muriquis?

Just then there was a loud swishing sound from the slope just below, and a long, loud whinny exactly like a horse's neigh. A small tree rebounded as a large golden-gray monkey leapt from it into another, and then swung gracefully from a thick lateral branch. A second later the monkey was out of view, hidden in the leaves. Another swish a few meters from the first, and a female with an infant clinging tightly to her back was in and out of view. Only the swaying of the treetops was visible as the muriquis swung across the valley and up the other side of the slope.

Mittermeier and I sprang to our feet and took off, scrambling, sometimes crab-like on all fours, down the steep trail. By the time we reached the spot where the muriquis had disappeared, there was no longer any sign of movement. The air was full of a sweet, spicy scent, which I later learned was characteristic of muriquis and was probably due to the cinnamon leaves that they eat. There were no trails in the direction the muriquis had gone, and it would have been impossible to try to track them so late in the day. Although I had seen my first muriquis for less than a minute, their vocalizations, locomotion, and smell were already imprinted in my mind.

During the next two months I saw the muriquis on several occasions. Most of these sightings were along the road and trails; first when they crossed from the *Jequitibá* trail to the Raphael side of the forest, and then when they crossed an old cattle path from Raphael into Sapo (Figure 3.5). While they were along the road it was possible to watch them without disturbing them, but they were still between 50 and 100 meters away, which was just close enough to see whether they were eating, but too far to see what type of food it was.

Seeing all of the 22 muriquis in the group at that time strung out across a slope gave me a better sense of the challenge it would be to study them. It would be essential to be able to get closer to them, but keeping track of the group when the muriquis were spread out like this would also require some sort of strategy. Would it be better to stay with just one or a few individuals at a time, or try to monitor what all of them were doing by walking back and forth between them? Which tactic would reduce the likelihood of losing them, and increase the data I was interesting in collecting? Experimenting with different methods of observa-

Figure 3.5 The Matão muriquis cross the dirt road using arboreal bridges (photo by K.B. Strier).

tion gave me a chance to develop an idea of what kind of sampling methods would be most effective under these conditions.

I began to notice individual differences in facial pigmentation and shape, and fur color. Being able to recognize individuals from their natural markings was very good fortune, because it meant that it would be possible to compare the behavior of specific individuals. A few of the females quickly became familiar to me, but identifying the rest would have to wait until the group was more fully habituated and could be watched at closer range.

RESEARCH DESIGN

I returned to the United States in August 1982 to develop my research plans with a clear idea of what could realistically be accomplished during a longer study period. The success of the study would depend upon habituating the monkeys so that they could be followed at close proximity without being disturbed or altering their behavior. Many of the questions I was interesting in addressing would require that all of the individuals in the group could be recognized. Habituating the muriquis would take time and patience, but the fact that they had not been hunted and were accustomed to seeing local farmers along the dirt road would make it easier. The challenge, however, would be to follow them away from the road, deep into the forest. There, once they allowed me to move closer to them, I would be able to compare the natural markings of each individual and learn to distinguish them.

Hypotheses

It was possible to generate alternative sets of hypotheses about muriqui behavior based on whether muriquis were folivorous—feeding primarily on leaves—or frugivorous—feeding primarily on fruits. All primates the size of muriquis include both fruits and leaves in their diet, but the relative proportions of these food types are correlated with a number of other behavioral features.[13] Activity levels, daily distances traveled, and home range size, or the total area used, generally increase with the proportion of fruit in the diet because fruits are more widely distributed than leaves. Folivorous primates, by contrast, tend to have shorter day ranges and smaller home ranges because of the more abundant distribution of leaves, and lower activity levels because leaves are comparatively low in energy.

Characterizing muriqui dietary preferences and feeding behavior was essential to testing predictions about their social organization. The 22 members of the Matão group in 1982 included both adult males and females, so comparisons with other primates could be restricted to those species with similar multimale, multifemale groups. Three alternative scenarios could be evaluated, depending on whether muriquis were specialized frugivores like spider monkeys and chimpanzees, able to shift their diet from fruit to leaves like the majority of Old World, female-bonded monkeys, or specialized folivores like howler monkeys and gorillas. Each of these three dietary regimes is associated with a different type of social organization in other primates, providing models for the social organization of muriquis.

If muriquis prefer fruit, they should range widely through the forest to seek dispersed fruit resources, and, like their spider monkey relatives and chimpanzees, the group should split up into smaller feeding parties in response to the size of their preferred fruit patches. The strongest social relationships should be those between males, which would be expected to remain in their natal groups. Females, by contrast, should avoid feeding competition by avoiding one another, interacting only on the rare occasions when they aggregate at large fruiting trees.

If, on the other hand, muriquis can shift their diets to abundant leaves when fruits are scarce, their ranging behavior and diet should directly correspond to the distribution and availability of these foods. They should travel farther when large fruit patches are scattered throughout the forest, and restrict their movements to conserve energy when they are forced to rely on leaves. Their social organization should then resemble that of other female-bonded species with similar diets. Females would remain in their natal groups, have strong affiliative relationships, and cooperate as a cohesive group in defending available fruit resources, while males would disperse and have strongly competitive relationships with one another over access to females.

Finally, if muriquis are strongly folivorous, they should conserve energy expenditure by using a relatively small area with just enough leaf sources to meet their nutritional needs, and devote a large proportion of their day to resting. Social relationships among both males and females should be weak, and both sexes should disperse from their natal groups because there are no advantages to either sex to cooperate with kin.

Descriptions of muriquis from museum specimens indicated that they would not fall neatly into any of these discrete behavioral categories.[14] Muriquis possess adaptations for feeding on both fruits and leaves, and integrating the anatomical reports with my preliminary observations from 1982 was equally paradoxical. Although the muriquis had often dispersed over a 100–300 meter area when they were feeding, they had also maintained vocal contact with one another and coordinated their movements so that when they traveled to another part of the forest they did so together. Cohesive social groups are characteristic of both female-bonded and folivorous primates, and although the majority of the feeding observations involved leaf-eating, there was little fruit available in the forest during the months when these data were collected. It wasn't at all clear whether muriquis eat leaves year-round, or only when fruit shortages force them to shift their diets. At the same time, the muriquis had traveled longer distances than they should have if they were restricted to a low-energy diet of leaves, but it was possible that they were simply fleeing from my unfamiliar presence rather than searching for fruit.

The few social interactions I had witnessed confounded the socio-ecological predictions even further. The adult males almost always appeared to be in close proximity to one another, yet there was no indication of any competitive tension between them. In fact, they occasionally embraced one another before settling down to feed or rest, suggesting that male muriquis maintain strong affiliative relationships similar to those among the frugivorous spider monkeys and chimpanzees. Why the muriquis ate so many leaves if they were really specialized frugivores made no sense from a comparative perspective, but it would be necessary to follow the muriquis over at least an annual cycle so that their responses to seasonal shifts in food availability could be monitored. Collecting the data that would permit me to evaluate how muriquis diverged from other primates would require systematic measures of diet, activities, ranging patterns, grouping patterns, and social relationships. Because diet was such an important variable, it would also be essential to collect independent data both on what foods muriquis ate and on what foods were potentially available to them. Their dietary preferences could then be compared to data on their activity patterns, ranging patterns, and social interactions.

Behavioral Sampling

I developed a research methodology that would be feasible given what I knew about the muriquis and the observation conditions in the forest. The majority of my data were collected using scan samples,[15] in which the location, activity, and nearest neighbors of all individuals in view were recorded at intervals of 15 minutes throughout the day. I used the trail measurements to construct a map of the forest, and superimposed a grid system onto this map. Each quadrat measured 100 meters by 100 meters, and was assigned a letter and number coordinate.[16] References for the muriquis' location would be much more precise than this when the animals were close to one of the measured trails, but when they were not near a trail I would need to estimate their position with less accuracy.

Their activities were divided into general categories and assigned a single-digit numerical code which could be appended with more specific information. For example, if an individual was feeding during a scan sample, it was recorded as "3." The second digit indicated the food type, so feeding on immature fruit was 31, mature fruit was 32, fruit of unknown maturity was 33, flower buds were 34, mature flowers were 35, immature leaves were 36, mature leaves were 37, leaves of unknown maturity were 38, seeds were 39, mature fruit and seeds were 329, and so on. This system, adapted to each broad category as new observations required new distinctions, enabled me to expand the original categories without modifying or losing any information, and to analyze my results in various ways depending on the questions being addressed. To determine the proportion of feeding individuals observed, all activities beginning with a "3" could be grouped and compared to other activity categories; to determine the distribution of food types eaten, all feeding observations on fruits and seeds (31, 32, 33, 329), flowers (34, 35), and leaves (36, 37, 38), could be analyzed.

Interindividual distances were important to understanding muriqui spatial relationships as well as social relationships. The distances between "nearest neighbors" were divided into five categories, which were also numerically coded: 0—in contact; 1—within a 1 meter radius; 2—within a 5 meter radius; 3—within a 10 meter radius; and 4—greater than 10 meters. The individual or individuals closest to the muriqui I was sampling at that moment could be recorded by name once I could recognize them, and I soon found that nearest neighbors were not always reciprocal.

Irv and Mark might be within 1 meter of one another, while Scruff was within 5 meters of both Irv and Mark. In this case, Irv was scored as Mark's nearest neighbor, Mark as Irv's nearest neighbor, and both Irv and Mark as Scruff's nearest neighbors. It was not clear to me at the time whether a distance of 1 or 5 meters meant anything to the muriquis themselves, but they were categories that could be reliably distinguished with ease. By analyzing the data separately, it would be possible to determine whether spatial relationships differed between individuals, and how their spacing related to their various activities.

The scan samples were the most effective method for collecting general data on activities, range use, and social relationships because they represented nearly instantaneous observations during the 5 second period allotted to record each individual. But in order to examine more dynamic social processes, I also conducted focal animal samples,[17] focusing on the adult males whose relationships appeared to be so unusually strong. A focal male was selected from a predetermined list, and its activity, nearest neighbors, and responsibility for moving into and out of proximity were recorded continuously. The focal animal samples could not be used to evaluate how much time muriquis spent in different activities because animals were always moving in and out of view. But when the muriquis were resting, the focal samples provided information about how important spatial relationships were maintained.

The systematic behavioral observations were also supplemented with opportunistic recordings of rare events. Sexual inspections, copulations, embraces, aggressive interactions, and intergroup encounters were defined and scored whenever they were observed.

Between the scan samples and focal animal samples, I was able to collect data on general behavioral patterns and the dynamics of social interactions. But what was missing was a way to evaluate the dynamics of feeding behavior. To do this, I developed a method that opportunistically focused on muriqui feeding trees.[18] In these "feeding tree focal samples," continuous records of each individual who entered a food tree were maintained until all individuals had left the tree, making it possible to examine whether muriquis fed together or sequentially, and whether there was any relationship between the size of the food trees and the number of individuals feeding together. The average and total time spent in a food tree could also be determined from these records.

Ecological Sampling

In order to evaluate the availability of potential muriqui food resources, I surveyed vegetation plots at four week intervals for evidence of fruiting, flowering, and new leafing events.[19] The location of the vegetation plots was determined by superimposing a numbered grid system onto a map of the part of the forest that the muriquis occupied. The first 20 squares whose coordinates coincided with the first 20 numbers listed in a random numbers table became the locations of the plots.[20]

Because the vegetation at Montes Claros is very heterogeneous, it was preferable to sample a larger number of small plots rather than a smaller number of large plots. Plots were roughly three-tenths of an acre, or 50 meters by 25 meters in size. Each plot was divided into four segments, with narrow trails cut along the perimeters and halfway along each axis to improve accessibility. Since muriquis use their prehensile tails to suspend themselves from sturdy supports while they feed in tiny trees, all trees and lianas 2 meters or more in height were potential food sources and needed to be sampled. The proportion of each tree and liana's canopy comprising fruits or flowers was estimated at <1%, <5%, <10%, <25%, <50%, <75%, or <100%. The diameter at breast height (DBH) of any tree with fruit or flowers present was measured, and the height from the base to the lower canopy and to the upper canopy, as well as the canopy shape, were estimated. Each of these plants was marked with plastic surveyor's flagging tape, and a numbered aluminum tag, and, whenever it was possible, botanical specimens were collected for subsequent identification. Once a tree or liana had been marked, its leaf production was also monitored for the remainder of the study.

These phenological data could be used to evaluate when fruits, flowers, and new leaves were most abundant in the forest (Figure 3.6). By weighting the proportion of the canopy containing these potential food types by the size of the canopy, I could also evaluate the relative availability of food types. Large trees with few potential food items may have been less important than smaller trees with a high density of potential food items, and I could examine these differences relative to what the muriquis were actually eating by collecting the same data on muriqui food sources that I collected on the plants in the vegetation plots.

I was also interested in whether muriqui feeding behavior and feeding group sizes varied in response to the size of food patches.[21] A food patch is generally defined as a tree or liana whose canopy is discontinuous with those of other members of the same species. The physical size of the patch is important, but measuring the canopies of all muriqui food sources would have taken valuable time away from studying the monkeys. Tree diameter is strongly related to canopy volume, and, if I could calculate this relationship for the vegetation at Montes Claros, I could use the more easily measured diameters to predict canopy volume.

About halfway into the study, after a number of muriqui food trees and productive trees from the vegetation plots had been marked, I chose 50 of these at random. The height of each canopy was measured with a

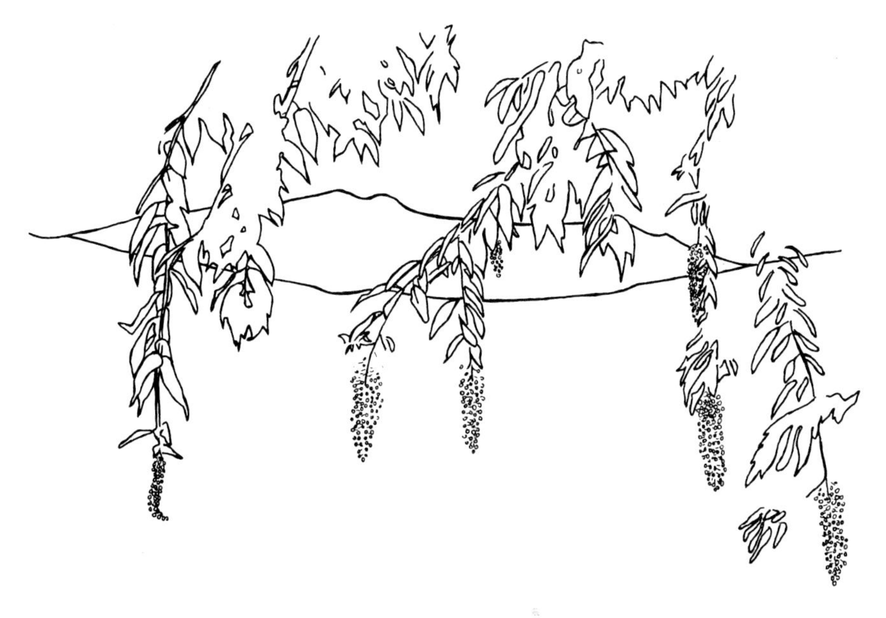

Figure 3.6 Flowers of *Mabea fistulifera* (Euphorbiaceae) are an important source of nectar for muriquis and other primates.

clinometer and the north–south and east–west projections of the canopy onto the ground were measured with tape. By inserting these measurements into the appropriate volumetric equation for the canopy's shape, I could calculate canopy volumes, and plot them against their trunk diameters.[22] This gave the relationship that I had hoped for, allowing me to estimate food patch sizes from the measured trunk diameters.

Arrangements were made to have all of the collected plant specimens identified by experienced botanists at the New York Botanical Garden and the Federal University of Minas Gerais. Each evening that I returned with plant samples, I would press them between newspapers and dry them for preservation. All loose fruit and leaf samples were also weighed before and after drying to determine their water content, and then saved for later biochemical analyses to determine the nutritional content of muriqui foods.

It took me months to develop my hypotheses and methods of data collection, and to write convincing grant proposals. But even while making these plans, I knew that conditions might unexpectedly change and that it was essential to retain some flexibility. The tension between rigorous planning and the inherent need for flexibility is exaggerated in field conditions (compared to laboratory conditions) because immediate feedback in the field is usually lacking. The data accumulate long before they can usually be analyzed, and the isolation of most field sites, including Montes Claros, from libraries or colleagues meant that there would be few sources to consult once I was back in the forest.

4

From Days to Years

I left the United States for my second visit to Fazenda Montes Claros in June 1983 with five large suitcases containing the clothes, field supplies, and books I would need for the next 14 months. Unfortunately, I arrived in Rio de Janeiro alone. I was promised that my luggage would be delivered to me within a day, which had its advantages since now I would not have to try to explain the purpose of each item to the customs officers in my still somewhat broken Portuguese. I had arranged to have Mittermeier's Brazilian assistant, Carlos Alberto, pick me up at the airport and drive me to the forest, and when my luggage was delivered the following morning, we set off.

Before leaving the United States, I had planned my first two weeks in the forest precisely, knowing that the new sensations, confusion, and loneliness would make it difficult to think on the spot. By jumping right in to my preformed plan, I suspected that I would adapt more quickly to the transition and to any unexpected surprises. My two months at the forest the previous year had prepared me for this extended stay, but sometimes the responsibility and reality of what I was about to do overwhelmed me. This was IT; I had just 14 months, and no second chance, to collect enough data to write a doctoral dissertation and to satisfy the organizations that had generously decided to fund me.[1] Fourteen months seemed like an eternity, but realistically I knew it was very little time. Everything would have to work, and, if it didn't, it was up to me to fix it.

When Carlos Alberto and I arrived at the farm, we greeted Sr. Feliciano and then drove the 2 kilometers to the small white stucco house at the edge of the forest that Sr. Feliciano had just donated, along with the 2.5 acre piece of land that it stood on, to the Brazilian Foundation for the Conservation of Nature (Figure 4.1).[2] I unpacked the rain gauge and the thermometer that I would use to monitor the climate of the forest, and Carlos Alberto helped me to set them up. We attached the rain gauge to a stake in an open area behind the house, where no trees would drip into it, and hung the thermometer from the side of a big tree about 5 meters into the forest along the closest trail, where it would be protected

Figure 4.1 The Research Station of Caratinga was donated by Sr. Feliciano and renovated by various conservation groups for researchers working in the forest at Fazenda Montes Claros.

from direct sunlight. Somehow, setting up my rudimentary meteorological station was a huge relief.

The house was full of people, but they spoke mostly Portuguese and when they sat around at night laughing and talking in a language I was still barely able to follow, it was incredibly lonely. The days were easier to get through because there was so much work to do. Rosa Lemos de Sá,[3] an English-speaking Brazilian who knew Celio, along with one of Sr. Feliciano's employees, helped me to measure out the vegetation plots with tape and compass, and together we cleared paths around their peripheries. The plots were distributed randomly through the forest, in diverse habitats ranging from dry ridge tops to moist valleys. They encompassed secondary vegetation in areas where the forest had been disturbed and was in the process of regenerating, and old primary vegetation where the forest had been left intact. Some plots were conveniently situated near existing trails, others required that we slash through underbrush until we reached the designated site.

When the plots were all set, Rosa and I began to search for the muriquis. Two other Brazilian students staying at the research house also worked in the forest: Gustavo Fonseca was collecting some data on muriquis for his Masters' thesis, and Sergio Mendes was studying the howler monkeys for his.[4] But by the end of the first month, both Rosa and Gustavo had left, and I was alone with Sergio. At night we practiced my Portuguese, but I spent the days, from sunrise to sunset, walking through the forest.

CLOSE ENCOUNTERS

When I crossed paths with the muriquis along the road I could observe them for hours, but deep within the forest, away from the road, they kept their distance and would flee when I tried to move closer. I had to habituate them, and learn how to recognize the rest of the group; my early observations did not allow me to distinguish individuals. I rarely spent an entire day with the muriquis, and often they were infuriatingly scarce for two or three days at a time. I tried not to let my frustration overwhelm me; things had to improve if I just kept at it. In July I saw that Mona, a female I remembered from the previous year, was carrying a newborn infant. I counted the same 22 individuals from 1982, and, with Mona's infant, the group now totaled 23.

One afternoon, after searching for the muriquis all morning, I came upon a group of them scattered in a cluster of trees. I heard their munching while they fed on leaves, and the occasional soft clucking of a mother calling her dawdling offspring to climb on her back before she swung to another spot. I decided to move closer to the group, clearing my throat so that they would not be surprised by my sudden arrival. As I drew near, three adult males suddenly turned toward me and rapidly approached. They were agitated as they took positions almost directly overhead. They began to bark loudly, alternating calls with one another, which created the impression that there were many more than three involved in this vocal display. They flailed their arms, balancing themselves on slightly bent legs with their tails securely grasping a branch behind them. They were almost as tall as me, and with their arms outstretched they looked even larger. They constantly touched and occasionally hugged one another, but they never took their eyes off me. Their deep calls and frenetic movements attracted other group members, who temporarily joined in the threats, but I kept my attention focused on the three males who had initiated this first interaction.

The three males were very close now, less than 2 meters above me. I had never been this close to them. Large branches were breaking; those with leaves floated safely to the ground but thicker barren boughs crashed menacingly nearby. I resisted the impulse to move out of danger, afraid of scaring the muriquis away. One medium-sized branch did strike my shoulder; I was sore and bruised for days. The muriquis were also defecating. The ground around me, by boots, and back, were splattered with their moist, leafy-green droppings which had an aromatic, unoffensively spicy smell. Slowly, I eased myself into a squatting position; maybe I would appear less intimidating if I made myself smaller.

This gesture seemed to help, although it is also possible that they were tired of keeping up the commotion. The males calmed down and hung suspended by their tails, the three furry forms intermittently grasping one another with a reassuring hand or foot, now and then stuffing a nearby leaf into their mouths in a distracted manner. They were studying me as carefully as I was studying them. They were beautiful, with their thick golden-gray fur and their cinnamon scent.

I wanted to examine their faces, to begin to look for the clues that would permit me to tell them apart and to recognize them individually. One had a deformed upper lip and a lot of pink pigmentation around his nose; another had a black face with a white cross between his nose and upper lip; while the third had lighter fur than the others, and a black face with no other obvious distinguishing features. Slowly, moving only one hand, I reached for my binoculars. They tensed again when I took aim, as if torn between curiosity and concern. Suddenly another male, who must have worked his way behind me during the earlier chaos, came crashing overhead, sounding the alarm to the three who had been watching my movements so intently. Just as he passed, a piece of dung struck the left lens of my binoculars and my hand; spontaneously I jerked my hand back, provoking them again. With considerable noise and shaking of branches, all four males swung off to join the rest of the group, which had already moved up the slope. I could hear a series of neighs that greeted their return.

I stood up, shaking my legs, which had grown quite stiff. I was out of breath, probably because I had been breathing very shallowly the entire time. I took off through the forest in the general direction the muriquis had gone, but I got stuck in an almost impenetrable patch of bamboo near the top of the ridge and my progress was slowed. For the first time, I forgot about the ever-present danger of poisonous snakes that might be hiding in the dense vegetation. There wasn't time to carefully check each hand-hold before pushing the bamboo aside and hoisting myself through. I was too excited by my first real encounter with the muriquis, and I wanted to catch up to them before dark.

It took forever to extract myself from the bamboo patch. By the time I reached the spot on the top of the ridge where I had last seen them moving, the group was gone. I waited as long as I could, scanning the folds of the hills and valleys below for any sign of movement in the trees. Night falls suddenly in the forest, and it becomes impossible to see more than an arm's length ahead. The light was already fading up here, and I knew it would be even darker as I descended the shadowy trail. Reluctantly, I conceded that I had lost the muriquis. Or rather, they had lost me.

I walked at least 10 hours a day for the next three days in search of the group, up and down the overgrown trails, which were always in need of repair. It was hard to hear above the sounds I made as I trampled dry

Figure 4.2 Muriquis often sleep in the upper canopies of large trees (photo by K.B. Strier).

leaves, so I devised a system which I still use: walk 10 paces—stop and count to 20 while surveying the treetops, listening to the forest, and sniffing the air for their cinnamon scent—walk 10 paces—stop for 20 . . .

I didn't see the group again until the third evening. I had just returned to the research house, and was sitting on the veranda with a cup of strong coffee left over from the morning thermos. As I sat listening to the night sounds while the fatigue of the day slowly engulfed me, I suddenly heard a familiar, horse-like "neiiigh" from the top of the ridge just in front of the house. I knew what this meant, and almost immediately I spotted the silhouette of a muriqui suspended against the dusky sky (Figure 4.2). I watched as it swung, hand-over-hand, into a tree where it would spend the night, and I wondered whether the group had been here, at my doorstep, all these days while I had been searching for them everywhere else in the forest. I ran to the dirt road to search out the trail entrance that I would climb at sunrise, plotting where I would have to leave it in order to reach them before they woke up.

Throughout the night I thought I heard their intermittent neighs, but I'm not sure whether they were real or the sounds of dreams. Given what I had been through the last three days, finding the monkeys camped just outside the house was a major accomplishment, and I was up, dressed, and ready to go long before dawn. A little overeager, I had to wait for the sun to rise so that I could find my way up the trail.

When I reached the muriquis, they were still asleep. A few awoke, neighed and looked at me, but they didn't move. I spent the next seven days with them, never losing them for more than an hour or two, and then only because they could travel more rapidly through the canopy than I could on the trails below them. They kept returning to feed at a large fig tree during this period, and I could observe individuals side by side for hours at a time. By leaving the trails to follow them wherever they went, I was less likely to lose the monkeys unless I got stuck in some dense vegetation. I gave up carrying a machete to hack my way through the forest because using it made so much noise that I couldn't hear where the group was moving. It was faster, and quieter, to scramble through the vines, bushes, and bamboo patches with only gloves to protect my hands.

After following the group continuously for a week, taking time out only to sleep, I reluctantly abandoned them in their sleeping area one evening. The next day I had to begin a week of working in the vegetation plots, and even if I heard or saw the muriquis during this time, I would not be able to follow them.

It took five days to survey the 20 small plots. The vegetation was still unfamiliar, and I had to find several spots in each plot where the canopy around me was fully visible. It was August, and there was very little fruit in the forest at this time. Apart from some large figs, there were long, bean-like pods of some legumes and some velvety orange fruits of a species, *Mabea fistulifera*, in the spurge family, Euphorbiaceae.

The assemblage of plants in each plot was unique. Plots on the upper slopes and ridge tops were open, with just a few species of short trees and very few lianas. Plots on the lush lower slopes and in the valleys were a tangle of species. At the lower altitudes, the canopy was much higher, and it was difficult to see whether any fruits or flowers were present 30 meters overhead. There were also more lianas, and it took a lot of time and concentration to trace their gnarled growth through the trunks and branches that supported them.

Working in the plots was boring and lonely. Somehow the lack of human company never seemed so oppressive when I was with the muriquis. But, after a couple of months, as I learned more, the plots began to seem more interesting, and I began to see the forest from the perspective of its fruiting and flowering cycles.

When I sighted some fruit or a flower in the canopy, I looked for representative branches or individual items among the detritus on the ground. Each night I pressed the branches I had collected between newspaper and weighed the individual items before setting my plant press and collecting sacs on top of the wood-burning stove to dry. I hadn't seen the muriquis since my second day in the plots, and I wondered where they had gone.

Once the phenological samples were complete, it took three days before I found the monkeys again, and it began to dawn on me that each month I would lose valuable time relocating the muriquis after finishing the vegetation work. I needed an assistant who could stay with the monkeys while I was devoting my attention to the plots. Another person would also be a big help when the muriquis disappeared; we could split up and search an area twice as large as I could search by myself. I relayed these thoughts to Celio when he dropped in for a two day visit to see how the work was progressing. He promised to look for an assistant for me among his students at the university. I assumed it would be months before Celio would be able to find someone interested in helping me, but a week later, Eduardo Veado appeared.

I was getting ready for dinner with Sergio, whose studies of the more sedentary howler monkeys were going quite well, when a clean-cut figure about my age walked up the road from Sr. Feliciano's farm with a backpack on his shoulders. He explained that he had been planning to take a year off from his university studies, when Celio suggested that he spend the year working with me. I was hesitant to promise him anything; after all, it is one thing to conceptualize the ideal assistant and another thing to rely on a stranger who has no prior field experience. In my improved, but still faltering Portuguese, I suggested that Eduardo spend a few days familiarizing himself with the forest. I wanted to be sure that he knew what he was getting himself into before I invested the time or energy in training him.

The next morning Eduardo climbed one of the trails to the top of the ridge in front of the research house. There he discovered an old trail leading down into the valley on the other side, and he followed it into the Jaó part of the forest. It was late in the day when he reached the dirt road that runs through the Jaó valley, so he wisely decided to make his way back along deserted country paths rather than try to negotiate overgrown trails in a darkening forest. Back at the house, Sergio and I were getting worried. Had Eduardo gotten lost, or, worse, hurt? We grabbed our flashlights and were just setting out to search for him when we saw him coming up the road. The next morning he was up and ready to explore another part of the forest, and I knew that Celio had found me an assistant.

Eduardo continued to work with me for the duration of my study, and accompanied me on visits to other muriqui field sites in subsequent years. In 1986 he became the director of the field station at Fazenda Montes Claros, helping to coordinate the muriqui research, other studies, and the large number of tourists, many of them Brazilian, who began visiting the site year-round. By 1988 he had graduated from the university, married, and moved to Sto. Antonio, a one-street town 10 kilometers from the forest, where he now lives with his family and administers the research station.

RAINY DAYS

Eduardo stayed with the muriquis whenever I had to focus my attention on the vegetation plots, and helped me to search for the monkeys whenever they ditched us. I had been in the forest for over three months. I could recognize all individuals in the group, and had assigned each of them a name based on physical characteristics or behavioral traits that reminded me of people I knew. The male I had distinguished by his deformed lip the first time the muriquis turned to threaten me became Cutlip, or "Leporinho" in Portuguese; an adult female was named Sylvia after a favorite aunt of mine.

I left the forest for two days to go to Belo Horizonte. I needed to check in at the U.S. Consulate, and already I needed a new pair of boots. I took the 4:00 pm bus that passed in front of the farm on its way to Caratinga, and waited 5½ hours in the open-air bus station there for the midnight bus to Belo Horizonte. Arriving in Belo Horizonte at 6:00 am the following morning, I walked the 3 kilometers to the Consulate. It was strange being in a city again. I kept forgetting to watch for traffic when I crossed the streets, but the sudden movement of any debris swept up in a gust of wind on the ground triggered an adrenalin rush as I automatically stopped to make sure it wasn't a snake. My eyes burned, unaccustomed to the automobile exhaust, and my ears hurt from the sharp city noises.

Only when I entered the Consulate, and its elegant surroundings, did I realize what a mess I must have looked. My best jeans and T-shirt were wrinkled and sloppy from the overnight bus ride, and my arms and hands were covered with cuts and swollen bug bites. I left the Consulate as soon as I had completed my business, made my various purchases, and went back to the bus station. I bought a ticket on the late night bus to Ipanema. Although it was a longer ride than the one to Caratinga, I could get into Ipanema in time to catch the 5:00 am bus that stopped in front of the farm at 6:00 am. Eduardo had left me a note telling me where the muriquis were, and I was back in the forest before mid-morning.

The days were getting longer with the onset of spring, but with Eduardo's help I never lost the muriquis for more than a day or two at a time. Keeping up with the group was exhausting work; the muriquis were traveling over 2 kilometers each day, more than twice as far as they had previously, and they were leading me to new parts of the forest which were not included in the trail system. When they left these areas, I asked Eduardo to open new trails and connect them with existing ones.

I had been on my own, in a foreign country speaking only Portuguese for four months, living in rustic conditions with few material comforts. Sergio and I had become good friends, and Nadir Ferreira,[5] the woman who cooked and cleaned for us, had become a trusted confidante. But sometimes the loneliness and nostalgia for my friends in the States made me question what I was doing there.

Letters helped allay these feelings of isolation, and I always took advantage of the monthly supply trip to Ipanema to call my family from the only accessible telephone in the region at the time. But during the rainy season, the bus to Ipanema couldn't negotiate the muddy road that passed by the farm. Being cut off from any access to my other life was not sufficient reason to risk a 26 kilometer walk, but, finally, we were so short on food that Eduardo and I caught a ride to Ipanema with the milkman, whose truck somehow still managed to make it through the mud. No vehicles passed us to hitch a ride on the way back, and we had to walk the full 26 kilometers, slipping and sliding with bags of food strapped to our backs.

As my contact with the animals increased, they introduced me to new species of food that they ate, and allowed me to witness new behaviors. In fact, the period between September and December of 1983 was full of first-time observations: muriquis playing on the ground; muriquis eating bamboo; muriquis falling out of trees. The group was noisier now than ever as the six juveniles born in 1982 screeched and whined while their mothers weaned them. The excitement of these discoveries and changes helped sustain my energy and enthusiasm; I knew I was collecting useful data, and I quit worrying about whether I would have enough to write my thesis.

Visitors began arriving toward the end of December, and they helped break up the monotonous routine. Cristina Alves, another of Celio's students, spent a week or so each month in the forest studying the marmosets, and soon she became a friend to stay with when I went to Belo Horizonte. Celio and Ilmar, the student who had helped me measure the trails the previous year, came for a few days every other month. I was always reinvigorated after Celio's visits, as he always made me feel that the study was important. In February, two new students from Belo Horizonte began a botanical survey of the forest. Although I had to share my tiny room with them for half of every month, it was nice having other women to talk to at night.

In March, the botanists brought news from Belo Horizonte that Mittermeier would be arriving at the end of the month with photographer Andy Young, some important Brazilian conservationists, and Celio's group from Belo Horizonte. This was the worst possible time for them to come: Eduardo had just gone back to Belo Horizonte for a few weeks' vacation, and I had lost the muriquis.

The day I heard the news of Mittermeier's visit, the muriquis had led me into Sapo, the valley at the southern edge of their home range. They climbed the southern most slope too late in the day for me to follow them, and they were nowhere to be found the next morning. I waited the entire day in Sapo, expecting to hear them at any moment, but by the end of the day I concluded that they must have crossed back into Raphael and the main part of the forest without my knowing it. By this time they could be anywhere.

As I searched for them during the next few days, I kept encountering muriquis from the Jaó group in the Matão group's home range. Sometimes I would mistakenly track these muriquis at a distance for hours, only to realize that I was following the wrong group when I finally caught up to them. Sergio listened for muriquis while he followed his howler monkeys. Nadir asked the local farmers who passed the research house each day whether they had heard or seen the muriquis along the road or in the forests that abutted their homes. Everyone was trying to help me, but after five days there was still no news. On the sixth morning, one of the farmers stopped me as I was leaving the house to tell me that he had heard some neighs behind Sapo. I wasn't sure which part of the forest he was referring to, but I went back to Sapo, and walked to the end of the trail where the forest gave way to a large field of sugar cane. The cane was at its full height, and I could barely see a thin patch of scrubby-looking forest along the ridge behind the field. I couldn't see any connecting vegetation that the monkeys could have passed through to reach this strip of forest. Disappointed, I realized that either I had misunderstood the information or that the well-intentioned farmer had confused the muriquis with some other animal. I returned to the trail, and followed it back through Sapo and into Raphael.

By the ninth day I was desperate. Mittermeier and his group would be arriving the next day. My fantasies of impressing him with how well I had habituated the monkeys and providing Andy with unprecedented photographic opportunities had completely evaporated. When Mittermeier and his group arrived, I stood before them, defeated, waiting for what I knew would be their first question "Where are the muriquis?" It was embarrassing to explain that I had last seen them 10 days ago, the longest period I had spent without the muriquis since my arrival.

I knew where the Jaó group was, and I brought Andy with me to photograph them. Mittermeier and the others met with Sr. Feliciano, while Carlos Alberto went to visit some of the farmers he had befriended. I had promised a case of beer if Carlos Alberto brought back any information that helped lead us to the muriquis, but I had little hope.

The sky was clouding over when Andy and I left the forest. As we approached the house, Carlos Alberto rushed out grinning, "*Cada minha cerveja*?" (Where's my beer?) We all piled into Mittermeier's van, and Carlos Alberto drove us to the sugar cane field behind Sapo. Instead of stopping at the entrance to the trail into Sapo, he veered up a narrower path in the middle of the field. Immediately I saw the muriquis, in the low strip of scrubby forest behind the field. The slope was too steep for the van to climb, and I jumped out and ran up to be sure these were the "right" muriquis. Just as I was able to recognize their familiar faces, they began to move, following what I now saw was a thin strip of mixed forest and bamboo than ran along the far perimeter of the field, all the way to the back of Sapo, and finally I understood where the muriquis had left me 10 days earlier.

They didn't return to this area, which we now call *A Mata de Dez Dias*, or "The Forest of Ten Days," for the rest of my study in 1984. However, it has subsequently become an important and much-used part of their home range; and as the forest has regenerated, even the Jaó group's males go there now.

EXPANDING THE RESEARCH

It was difficult to leave the muriquis and the forest after 14 months, but I was eager to return to the States and begin to analyze the data I had collected. So, on a crisp winter day in the middle of July 1984, I said good-bye to the muriquis who had become my companions and had allowed me to follow them for more than a year of their lives.

Although I had enough data to begin to evaluate my original hypotheses, I also knew that I had only begun to understand these monkeys. I had succeeded in habituating them and recognizing them individually, but new questions were puzzling me. Two subadult females from the Jaó group had joined the Matão group, suggesting that dispersal in muriquis might be characterized by female migration and male residency, but without further information I could not be sure. All six of the infants born in 1982 had been weaned during my 1983–1984 study period, and their mothers had copulated during the year, but only the female named Nancy had given birth. Had any of the other females also conceived? What would happen to Nancy's previous offspring, Nilo, now that he had been displaced by a younger sister?

Completely identifying and collecting many of the important plant species in the forest had eluded me, and, counting my first trip in 1982, my three years of observations during the months of June and July indicated that some important muriqui food species do not produce fruit annually. How would a year with low fruit availability affect muriqui diet and grouping patterns? I had seen numerous seeds in the muriquis' feces, and the majority were intact after having passed through the muriquis' digestive tract. How important were muriquis in dispersing these fruit species?

I had distinguished 10 different vocalizations and their contexts, but I knew that there were much finer distinctions to be made. Understanding the function of these calls and the development of vocal communication in immature muriquis would require more time than my study encompassed.

Finally, the relationships between the Matão muriquis and the Jaó muriquis were still a muddle. Most of the encounters between the two groups were antagonistic, with adults of both sexes threatening each other until one group or the other retreated. At other times, however, the two groups fed in adjacent trees, exchanging neighs, but no threats. Their relationships seemed ambiguous and highly variable, and I could not explain why.

Three months after I had returned to the States, in October 1984, I received a letter from Nadir. She apologized for not having written sooner, but a month after I left she had moved to Caratinga. She told me that the muriquis had appeared in the large plum tree in front of the research house just days after my departure, and that two different females were carrying tiny infants. I was certain that one of these females must have been Nancy with the newborn infant I had seen her with in June, but the other new mother could have been any of the females who had copulated the previous year. I wondered which one.

Three months later, in January 1985, I received a letter from Jairo Gomes, the man who had taken care of the research house the previous year.[6] He thought he had seen more than two females accompanied by offspring when he was clearing a trail. If he was right, at least two, and possibly three, other females had given birth since my departure. I realized that I could not ignore these two letters, which called me back to the forest before I had completed my dissertation. I needed to know what had happened.

Stepping off the bus at the farm in September 1985 for a two week visit, I was met by Sr. Feliciano and several of the local people. Despite my reassurances the previous year, none of us had really believed I would return so soon.

Jairo helped me find the muriquis my first day back. I immediately saw Nancy carrying what was now a 15 month old juvenile. There was Bess, also carrying a juvenile almost identical in size to Nancy's. Bess must have been the second female that Nadir had seen right after I'd gone. Both Arlene and Robin were carrying much smaller infants, the ones Jairo had written me about.

Apart from these births, the Matão group had remained intact. There

Figure 4.3 Sony, a subadult male, in 1985. This photo was taken with a 50 mm lens, showing how close and undisturbed the muriquis were by my presence, even after a 13 month absence (photo by K.B. Strier).

were now 29 individuals. They were still fully habituated, and I could still recognize all of them (Figure 4.3). The six infants I had first known in 1982 were large, fully-independent $3\frac{1}{2}$ year old juveniles. Four of them now had younger siblings, opening up the tantalizing prospects of examining the dynamics of kinship among muriquis. There were still so many questions left, but, unfortunately, I couldn't stay to answer them.

The night before my departure, a car pulled up to the research house. I was alone, and I had already locked up, but as soon as the driver mentioned Celio's name, I opened the door and let Dida Mendes in. Dida was a student from the University of São Paulo, who had just been to Belo Horizonte to talk to Celio about studying muriquis. Celio told him I was at the forest for a few days, and Dida had come to meet me. Within minutes of his arrival, I knew Dida was the solution to the long-term study I had begun to conceive.

We arranged to meet back in the forest in June 1986, when, in just a few weeks, I taught him to recognize the individuals in the group so that his observations would complement my own. It was my goal to have trained students in the forest continuously so that we could begin to answer some of the many questions left over from my first study. When Dida completed his research the following year, another student would take his place, and so on. Since 1986, the muriquis have been followed almost daily. Zé and Adriana Rímoli stayed with the muriquis from 1987

Figure 4.4 Each of the muriquis has distinct facial markings and fur patterns. (a) Adult female Louise. (b) Adult female Cher. (c) Juvenile female Cecilia, daughter of Cher (photo (a) by J. Rímoli; photos (b and c) by A.O. Rímoli).

until 1990, when Dida returned for his doctoral research. In 1991, he was replaced by Paulo Coutinho and Fernanda Neri, who are still with the muriquis now. All of the students develop original, independent studies, and contribute to the long-term demographic data by noting births, deaths, and migrations involving members of the group.

To facilitate the long-term record keeping, I adopted a naming system employed by many other field primatologists whose study subjects can also be easily recognized by their natural markings.[7] I began by assigning names to all of the independent muriquis present, then I followed the rule that the names given to dependent offspring must begin with the same initial as that of their mothers. Immigrant females could be given any unused name.

Whichever person is accompanying the group when a female immigrates or an infant is born is responsible for assigning a new name using the first initial rule. Thus, Nancy's infant from 1982 was named Nilo; her daughter from 1983 was named Nina; her son from 1987 is Nelson; and her daughter from 1990 is Nadir. Bess' offspring, in order of birth, are Bruna, Brigitte, Beatrice, and Bernardo.

Using the first initial rule helps us to keep track of matrilineal kinship. Adult and subadult muriquis at this site are easy to identify because they have distinct facial markings that help to distinguish them (Figure 4.4), but infants and juveniles are more difficult to identify because until about four years of age they all have similar black faces. For younger animals we must rely on more subtle differences, such as fur coloration, ear position, and brow and nostril shape. Even then, once their faces begin to change it is often difficult to recognize them after a long absence, and I am increasingly dependent on the students who follow these changes on a daily basis to help me keep track of maturing individuals each time I return.

5

Early Risers and Other Surprises

The summer sun rises at 5:00 am in southeastern Brazil, and the forest begins to wake up like a vast green blanket unrolling with life. It is already humid, and dew steams from the tops of the trees. Troops of howler monkeys fill the air with their long morning calls. Less resonant, the neighs of the muriquis signal that it's time for them to move on.

Muriquis spend the early hours of the long summer days traveling through the forest in search of food. But as the temperature rises to nearly 95°F around midday, the monkeys take refuge in the canopy's shade. Only when it begins to cool off in the late afternoon do they resume their foraging activities, often moving to another part of the forest, feeding along the way before settling down again at dusk, around 7:30 pm.

On days like these, I had to be at the muriquis' sleeping trees within half an hour of sunrise, and stay with them until sunset, 14 hours later. If I arrived too late, the muriquis would have already left their sleeping area and, not knowing where they had gone, I might have to search for them for days. Even when they stopped during the midday heat, I never knew how long they would sleep and I had to stay with them, resting myself before they led me on the inevitable afternoon trek.

In the winter, by contrast, temperatures fall as low as 45°F, and the muriqui's behavior is strikingly different. The sun rises at 6:30 am, yet the muriquis stay at their sleeping sites until mid-morning.[1] They move from the large, secure branches of the main canopy, where they spend the night, to the thinner branches in the upper canopy, where they sun themselves for hours before they start to search for food. By 5:00 pm, it cools down again and within an hour, just before sunset, the muriquis are already asleep.

Seasonal changes in temperature cause the muriquis to shift their schedule of activities, from mornings and evenings in the summer to midday in the winter, but the actual amount of time they spend feeding (19%) and traveling between food sources (29%) remains remarkably constant year-round.[2] But rainfall is also strongly seasonal in most of the muriqui's Atlantic forest habitats, and it leads to changes in the monkeys' behavior.

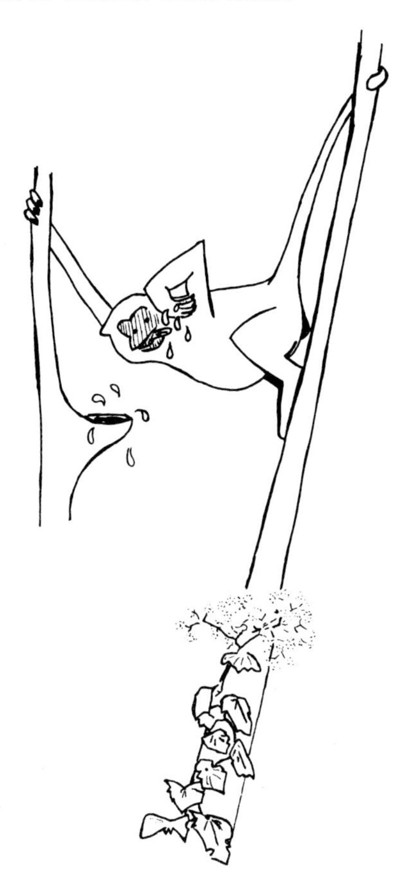

Figure 5.1 Muriquis drink water collected in tree cavities during the dry season.

At Fazenda Montes Claros, nearly 48 inches of rain fall annually, but 80% of this falls between November and April. The summer months, from December to February, are hot and wet, while the winter months, from June through August, are dry and cool, and many trees lose their leaves. In late winter, after months without rain, the only reliable source of drinking water is found in tree cavities. To get it, muriquis stick a furry hand into the cavity to soak the water up. Then, they pull out their hand, raise it over their head, and lick the drops from the outer side of their wrist (Figure 5.1). Once, toward the end of the dry season when even the tree cavities were empty, two muriquis came to the ground. They turned over rocks and fall branches until they uncovered a small puddle near a stream where they drank. In the rainy months, the narrow creak beds that run through the forest's valleys fill up, and the monkeys descend to the ground to lap water from these seasonal sources.

Figure 5.2 Myrtaceae berries provide an abundant source of energy during December (photo to K.B. Strier).

Drinking is not a daily activity, because muriquis usually get plenty of water from the plants they consume. But, rain or shine, they have to eat. Sometimes they stop to seek shelter during particularly heavy downpours, but when it rains all day long, they have to move on in search of a meal. Documenting the muriquis' behavior year-round meant that I had to follow them through the rain, slipping and sliding up and down the muddy trails, writing with my hand and my notebook stuck inside a plastic bag.

Seasonal rainfall also leads to seasonal cycles of fruit, flower, and leaf production in the forest, and muriquis have to shift their diets to accompany these cycles. Overall, muriquis spend about half of their feeding time eating leaves (51%), about a third eating fruits and seeds (32%), and about a tenth eating flowers and flower products, such as pollen and nectar (11%). The rest of their diet is made up of small quantities of bark, bamboo, ferns, and grasses, which they feed on from on the ground.[3] However, the importance of the various foods in their diet differs remarkably from month to month. During November, for example, nearly 80% of their feeding time is devoted to new leaf growth, but during December, 65% of it is devoted to fruits and berries (Figure 5.2).

Muriquis eat fruits whenever edible species are sufficiently abundant. Yet, they often ignore small concentrations of all but the most delectable fruit, perhaps because it is much more productive to keep looking for large concentrations where they can stop for a feast. But even in times of abundance, there never seems to be enough fruit to satisfy these large-bodied primates, so muriquis must always include some leaves in their diet. In fact, the necessity of relying on leaves to tide them over in times of fruit scarcity in the Atlantic forest may be the reason why muriquis are so large in the first place.[4] Leaves also provide essential amino acids that fruits lack, making them an important dietary supplement throughout the year. But although they are always included in the muriquis' diet, the ubiquitous leaves rise and fall in importance depending on the abundance and availability of more preferred food types.

Flower consumption is even more tightly linked to its availability than fruit, perhaps because flower production is so tightly synchronized in most plant species. The phenological data from the vegetation plots indicate three peaks of flowering at Fazenda Montes Claros: in October, when the delicate buds of *Apuleia praecox*, a member of the bean family (Leguminosae) appear; in February, when the bright yellow flowers of *Adenocalymna marginatum*, a type of trumpet vine (Bignoniaceae), burst open; and in early April, when the reddish spurge, *Mabea fistulifera* (Euphorbiaceae), is in bloom. Not surprisingly, the proportion of flowers in the muriquis' diet increases when flowers are abundant. The monkeys eat the entire mature flower of both *Apuleia* and *Adenocalymna*, but only the nectar and pollen of *Mabea*. By imitating the muriquis' actions with *Mabea* flowers that had dropped to the ground, I realized that when muriquis slide their mouths along the length of the flower, they stimulate the release of a deliciously sweet nectar. At the same time, their faces and hands become covered with the thick yellow pollen. Sometimes they lick their hands afterwards, but usually they are still coated in pollen when they move to the next flower. In this way, muriquis may inadvertently pollinate this species, contributing to its reproduction, just as its energy-rich nectar contributes to theirs.[5]

Energy from fruits and flowers, and protein from leaves, are the principal constituents of the muriqui's diet. But the forest is full of plants that produce chemical and physical deterrents to avoid being eaten,[6] and muriquis must distinguish between edible and inedible ones. Mature leaves of most woody plants contain high levels of tannins, which bind with the leaf's proteins and make it difficult to digest. The presence of tannins, as well as alkaloids and other secondary compounds, probably explains why muriquis rarely feed on any particular leaf species for any length of time. Several individuals will feed in sequence from the same tree, but each one feeds only briefly before moving to another food source.[7] Immature leaves, however, generally contain fewer tannins, and muriquis spend more time eating larger quantities of these when they are available.[8]

Figure 5.3 Muriquis prefer large fruit trees where they can feed together and camp out for days.

Unripe fruits tend to have more toxic compounds than the ripe fruits that muriquis prefer. When ripe fruit is plentiful in a single large tree, or food patch, the number of muriquis that feed together and the time they spend eating increases with the size of the patch. Unlike mature leaves, which must be eaten in only small quantities at a time, mature fruits can be eaten until the monkeys get bored or the food runs out (Figure 5.3).

Even seeds contain toxins, or are wrapped in thick coats for protection from predators. Muriquis chew small seeds along with the fruit's flesh, but larger seeds are often dropped after the flesh has been removed. Some seeds, however, are covered with a smooth, hard coat, a design that makes them easy to swallow along with the fruit. These seeds glide through the muriquis' digestive tracts, and appear intact in their feces (Figure 5.4).

Muriquis serve an important ecological function in dispersing these large, smooth seeds. Because they usually leave a fruit tree before their meal has been digested,[9] the muriquis effectively transport the seeds to another part of the forest, away from the shade and competitive environment of the mother tree. When my students collected seeds from muriqui feces and planted them elsewhere, they almost always germinated.[10] In fact, some muriqui-dispersed seeds actually germinate faster after they have been eaten than they do without the benefit of having passed through the muriquis' digestive system.

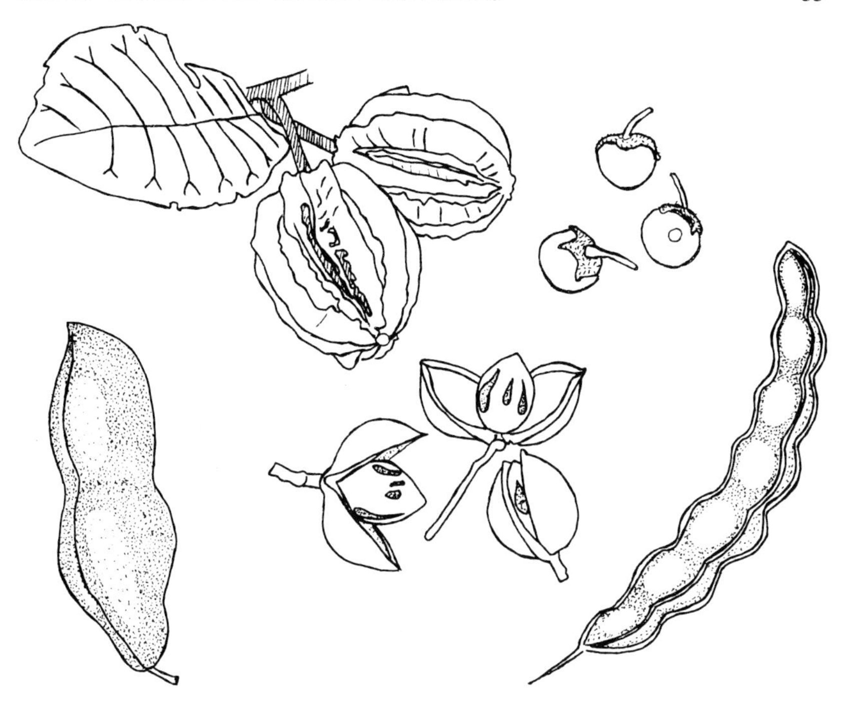

Figure 5.4 Muriquis disperse a number of seeds, either by dropping them or by swallowing them without chewing. Clockwise from top: *Sapucainha* (Flacourticeae), *Ameixa-bicha* (Chrysobalanaceae), *Ingá* (Leguminosae), *Bicuiba* (Myristicaceae), *Jatobá* (Leguminosae).

Muriquis eat only a few species of seeds that are not surrounded by fleshy fruit. The seeds of one of these, *Mucuna* (Leguminosae), are contained in large pods which are covered with a hard shell bristling with minuscule spines that break off and implant themselves in any hand or mouth that touches them. Adult and subadult muriquis are able to open these pods and eat the large, fatty seeds, but smaller muriquis avoid *Mucuna*, perhaps because they are not strong enough to open the pods, or because the fleshy pads of their fingers are not tough enough to deflect the spines. The *Mucuna*'s spiny defenses protect it from juvenile muriquis and other potential predators, but its fatty seeds are too important for larger muriquis to resist.

Leaves from the fig-like *Cecropia* (Moraceae) are another example of a food that muriquis go to great lengths to get. *Cecropia* have co-evolved with ants, which live in their trunks and defend them by attacking and stinging any animal that tries to enter the trees to feed.[11] To get around this defense, muriquis eat *Cecropia* leaves by suspending themselves from neighboring trees. I once witnessed an adult male attempting to reach a *Cecropia* from a small palm tree growing next to it. He rocked back and forth in the crown of the palm causing it to swing. Finally, the momentum of his weight brought the palm close enough to the *Cecropia* for him to grab a leaf before the palm swung back. He then sat in the center of the still-swaying palm tree to eat the leaf, his reward for a 10 minute effort. Biochemical analyses indicate that these *Cecropia* are especially high in protein and low in other toxic compounds, so the nutritional pay-offs appear to compensate for the time and ingenuity required to outsmart the tree's defense system.

The trade-off between nutritional gains and plant defenses may not be the only basis for muriqui food choices. Some of the secondary compounds that plants produce may also have important medicinal properties. Many indigenous human populations throughout the tropics exploit different plants and plant parts for pharmaceutical purposes,[12] and chimpanzees and baboons are known to eat certain plants for their antiparasitic and antibacterial compounds.[13] I suspected that the muriquis at Fazenda Montes Claros might find similarly important compounds in some of their foods, so in 1989 I began a study of the parasites in their gastrointestinal tracts. I relied on the information that could be gleaned from noninvasive examination of their feces, which were conducted by Dr. Michael Stuart and his students at the University of North Carolina-Asheville. Looking at over 80 fecal samples collected from 32 different monkeys, the results were astonishing.[14] The muriquis' feces were completely devoid of parasites, and our comparative analyses of howler monkey feces at this site show that they, too, are parasite-free. Most other primates, including howler monkeys and muriquis at other forests, carry in their gastrointestinal tracts a number of parasites that are passed along in their feces, so the fact that neither of these primates at Fazenda Montes Claros are infected suggests that something unusual is going on.

Parasites have evolved complex ecological relationships with their hosts, and the absence of parasites in both muriquis and howlers could mean that these ecological—and evolutionary—relationships have been disrupted at this forest. Perhaps a key secondary host has become locally extinct. On the other hand, it is possible that both primate species at Fazenda Montes Claros have discovered antiparasitic agents in their diets. Muriquis and howler monkeys eat many of the same plants, and by analyzing the ones that are found only at this forest we may be able to identify bioactive plants with value, not only for monkeys, but for humans as well.[15]

While medicinal properties may influence some of the muriquis' food choices, balancing their daily energy and nutrient requirements against the energy used in finding food is what governs most of a muriqui's day. When they aren't feeding or traveling to find food, muriquis conserve energy by doing nothing. In this respect, muriquis are no different from many other primates, which spend roughly 50% of the daylight hours resting between meals.[16] But the strong seasonality in food availability in the Atlantic forest means that the energy muriquis get from their foods and the effort required to get it also vary at different times of the year.

When fruits are scarce and the muriquis are feeding mainly on abundant leaves, they do not need to travel very far. On an average winter day, muriquis travel as little as 960 meters, just over half a mile. But in the summer, when fruit is plentiful, their average "day ranges" expand to over 1,400 meters, nearly 50% more than in winter.[17] In the summer, muriquis cover these longer distances without increasing the amount of time spent traveling by moving faster. The average muriqui pace in the summer is 1 kilometer per hour, but this is misleading since their movements actually occur in explosive bursts at speeds that may be four to five times faster.

Suspensory locomotion enables muriquis to travel rapidly. In fact, even after they were fully habituated to my presence, it could still be a challenge to keep up with them.[18] Sometimes they would smoothly make their way up a difficult slope, and by the time I reached the crest I could see the last of the group disappearing over the top of the next ridge. I learned to react to the first signs that the group was about to move, and I would follow the first animals when they began to leave. Then, even if they got ahead of me, I would still be accompanying the ones bringing up the rear. In this way, I was able to avoid losing the group every time they began to travel.

When moving quickly, the muriquis generally travel in a single file, using the same hand and tail holds as the individual ahead of them. Eventually I became familiar with particular travel routes, for when moving from one area to another, the monkeys frequently follow the same paths. It seems that their practice of using well-worn pathways is a strategy for reducing the chances of falling.

But, as adept as they are, muriquis are heavier than most other primates, and they do fall out of trees. On at least a dozen occasions during my first year with the muriquis, I saw animals fall from as high as 15 meters when the branch they had reached for snapped under their weight (Figure 5.5). Sometimes the muriquis were able to catch themselves by grabbing other branches as they fell; at other times, they hit the ground. When a muriqui lands, whether on the ground or in the vegetation, it scurries up to a larger, more secure branch and sits still for a few moments. Often, the animals seemed more stunned than hurt, especially if they had been moving slowly when they fell. Vegetation below them probably helps to cushion the impact, but in a few cases, muriquis appear to favor an arm or a leg days after a fall.

Figure 5.5 Muriquis are heavy primates, and avoid dangerous falls by following each other on well-tested branches when they are moving rapidly.

Despite the risks of falling, the muriquis' rapid locomotion is a big help in escaping from potential danger, as they did from me in the very beginning, and in traveling the long distances between widely scattered fruit trees.[19] Their great mobility enables muriquis to monitor a larger area for food than the slower, more sedentary howler monkeys at Fazenda Montes Claros. Consequently, even though they are larger than howlers and should need to eat more abundant leaves, muriquis can be pickier about their food.

Muriquis include more fruit in their diets than howler monkeys do, largely because they can find it more easily.[20] In this respect, the dietary differences between muriquis and howler monkeys in the Atlantic forest parallel those between spider monkeys, which also travel by suspensory locomotion, and howler monkeys in more equatorial forests.[21] The only difference is that both of the Atlantic forest species have to eat proportionately more leaves than their equatorial relatives because fruit supplies in the Atlantic forest are more seasonal.

The disadvantage of traveling over large areas means that muriquis may not find a particularly choice fruit tree until after a local howler monkey troop has eaten all the fruit. But if muriquis get to a tree while the howlers are still feeding, the howler monkeys usually abandon their meal, moving to an adjacent tree where they wait for the muriquis to finish and move off. If a howler is too slow to leave, or tries to ignore the new arrivals, the muriquis will chase it out of the tree. When a particularly large fig tree is laden with fruit, it is not uncommmon to see troops of both howler monkeys and capuchin monkeys sitting idly along the side while the muriquis eat. The smaller rivals swarm into the tree as soon as the last muriqui has departed, but, by then, there may be very little fruit left.

When muriquis reach a new area of the forest with abundant fruit, they typically remain there for several days, a pattern I call "camping out." Often, the monkeys arrive in an area a few days before most of the fruits have ripened. They hang around—quite literally—in the area, eating leaves and sampling the fruit as it matures, making only short forays into surrounding areas while they wait for the fruit to mature. When the fruit is ready, the muriquis gorge themselves and the tree is nearly depleted after a few days of this greedy banquet.

Afterwards, they may return to the part of the forest they had come from, or move quickly to another area with abundant fruit, where they camp out again. During periods when few large trees are bearing fruit, the muriquis scout around, sampling food as they meander through the forest. When scouting, the muriquis do not travel purposefully, in the rapid, single-file way that they do when moving between camp sites. Rather, they spread out in different directions, maintaining vocal contact with one another while out of sight. The animals come together only when they are about to settle down for a rest, or when one of them encounters a rare or particularly interesting patch of food, and alerts the others with excited chirps.

Camping and scouting for food in 1983–1984 led the Matão muriquis through an area of 420 acres, or roughly 20% of the forest at Fazenda Montes Claros. Due to the heterogeneity of habitats at Montes Claros, the group's "home range" encompasses various types of vegetation, from virgin primary forest, to regenerating vegetation in areas that have been selectively logged, to scrub and bamboo patches along the dry upper slopes and ridge tops. Much of the forest used by the muriquis is undisturbed, but they also spend considerable amounts of time in the regenerating and scrub habitats.[22]

Moving back and forth between these different areas enables muriquis to take advantage of the different plant species, with their own unique fruiting and flowering cycles, that grow in each area. In June, the muriquis camped out repeatedly in a valley where the mango-like fruits of an uncommon species, *Spondius dulci* (Anacardiaceae), were ripening. For the months of September and October, they meandered through the regenerating valleys where the *Apuleia praecox* flowers were in bloom. With the beginning of summer in December, the muriquis moved to the northern ridge, where abundant myrtle (Myrtaceae) berries were in season. Late April found the muriquis concentrated on the dry, upper slopes in the central part of their home ranges feeding on the plentiful *Mabea* flowers. They also fed opportunistically on fresh bamboo shoots and low-growing ferns as they crossed from slope to slope. In areas like the dry upper slopes, where the vegetation is especially sparce, the muriquis will even come to the ground to travel or feed.

At first I was surprised when the muriquis moved into the regenerating and scrub habitats because reports dating back to the 1800s emphasized that muriquis were found almost exclusively in primary growth.[23] Yet, over the years it has become increasingly clear that, at some times of the year, muriquis at Fazenda Montes Claros rely on foods found only in secondary habitats. A different picture of muriqui ecology emerges, however, at larger, less disturbed sites. Muriqui densities, or the numbers of individuals per acre, are lower at the larger, more pristine forests than they are at the smaller, more disturbed forests such as Fazenda Montes Claros.[24] While hunting and poaching pressure might explain the low population sizes at the larger forests, it is also possible that differences in food availability are responsible. The greater heterogeneity in vegetation and habitat types resulting from past disturbances at the small, private forests may actually provide more food to feed a larger number of muriquis. Consequently, understanding muriqui preferences for distinct habitat types is critical to interpreting their ecology.

Habitat differences at different forests may affect more than just muriqui population densities. At Fazenda Barreiro Rico, in São Paulo, muriquis are much more folivorous than those at Montes Claros. Based on observations collected during an 11 month study, Katharine Milton reported that muriquis at this site spend more time resting, travel less and in a smaller area, and live in smaller, more fluid groups than those at Montes Claros.[25] The Barreiro Rico muriquis' greater reliance on leaves probably reflects the greater scarcity of fruit in their forest, but why these muriquis live in such different kinds of groups is still difficult to explain.

In 1985 I visited Fazenda Barreiro Rico with two Brazilian colleagues in an effort to gain some insights into this question.[26] What immediately struck us was the absence of large trees poking above the main canopy in this forest, so we decided to make the most of our three day visit by measuring some trees in addition to looking for muriquis. The terrain at Barreiro Rico is flat, and although this makes it much easier to walk through than Montes Claros, it also means that there are no slopes to climb for a better view of the forest or the monkeys. On the first day we saw just three muriquis—two adult males and one adult female; the second day we saw only two adults, but they were too far away for us to determine their sex. On the third day we conducted what is known as a "plotless quadrant sample" to test whether our initial impressions of tree sizes at this forest were correct.[27] Following a 1 kilometer transect line, we stopped at 25 meter intervals to survey the vegetation. At each of these stopping points, we drew a line through the leaf litter perpendicular to the transect line, creating four imaginary quadrants, or quarters. We then measured the distance from the center point to the nearest tree in each quarter, the tree's diameter at breast height (DBH), and the tree's height. This offered a general estimate of the forest structure, which we could compare to similar measurements from Montes Claros.

Just as we thought, large trees were absent from this part of the forest. We had found no trees greater than 55 centimeters in trunk diameter along our transect.[28] While trees of this size make up less than 2% of the forest at Montes Claros, the muriquis there feed preferentially from these large trees, especially when they are full of fruit. Because trunk diameter is a good indicator of canopy volume, we speculated that muriquis at Barreiro Rico travel in such small, fluid groups because there are no trees big enough to support larger feeding groups.

Unfortunately, our ability to solve the puzzle posed by Milton's observations was limited because the part of the forest where Milton's muriquis lived had been sold to a neighboring farmer by the time of our visit, and our tree measurements were restricted to a connecting patch of forest still owned by our host. When I decided to continue the long-term research at Montes Claros in 1986, I began to wish that Milton had also continued her study at Barreiro Rico because it would have been interesting to compare our results. By this time, however, another comparative study was already under way at yet a third private forest, Fazenda Esmeralda, in Minas Gerais.

Six months into her 12 month study there, Rosa Lemos de Sá invited me for a visit. Fazenda Esmeralda is separated from Montes Claros by a few hundred kilometers of farmland, pasture, and highway, and its proximity to Montes Claros makes the climate, rainfall, and vegetation at the two forests quite similar. Perhaps because of these ecological similarities, the muriquis at Fazenda Esmeralda behave much more like those at Montes Claros than those at Barreiro Rico.[29] The diets of the two populations are similar in their proportion of fruits and leaves, and the 14 males and females in the group at Esmeralda travel together as a cohesive unit. Large trees are rarer in this forest than at Montes Claros, probably because it has suffered more from selective logging. But when

Rosa and I compared the size of muriqui feeding trees at the two sites, we found that both muriqui groups fed selectively at the few large trees available to them.[30]

Comparing the diets and grouping patterns of muriquis at Fazenda Montes Claros with those of muriquis at Fazenda Barreiro Rico and Fazenda Esmeralda has helped to illustrate some of the ways in which ecological variations, such as food seasonality and forest structure, can influence muriqui behavior. But both Fazenda Esmeralda and Barreiro Rico, like Montes Claros, are relatively small remnants compared to the original expansive Atlantic forest, and the vegetation has been partially disturbed by humans. What are really needed are data from larger, undisturbed tracts of forest that resemble more closely the areas muriquis originally inhabited.

Consequently, in 1988 I began two new studies in larger, ecologically distinct forests: the Augusto Ruschi Biological Reserve, located near Vitória, the capital of Espírito Santos; and Carlos Botelho State Park, only a $2\frac{1}{2}$ hour drive from São Paulo, the largest city in Brazil.[31] The two forests are virtually at opposite poles of the muriquis' current distribution, whereas Fazenda Montes Claros is closer to the center.

Unfortunately, the Augusto Ruschi Biological Reserve was a frustrating site for the students who worked there.[32] Although they regularly encountered marmosets, titi monkeys, capuchins, and howlers in the 2,000 acre area they searched, they spotted muriquis on only four occasions. In each case these were brief encounters, for the muriquis disappeared as soon as they detected the students. During the course of their census, the students also found evidence of hunters within the Reserve. Branches strung together in makeshift shelters and poles for drying meat were irrefutable signs that poachers had recently been through the area. The students notified the Reserve's administration, but with only five guards to patrol more than 11,000 acres of forest, it was not surprising that no poachers were ever caught. After a year, we reluctantly abandoned the study at Augusto Ruschi feeling that efforts to find and habituate a group of muriquis would be too risky to the muriquis, whose only defense from illegal hunters is to flee.

The other project, at Carlos Botelho State Park, is still under way despite different obstacles. The 2,000 acre area where we work is but a tiny fraction of the extensive forest, and when the muriquis move away from our trail system, deep into the forest, concerns about getting lost sometimes prevent us from following them. Nonetheless, the students who have worked at this site over the past four years have had increasingly frequent encounters with a group of 23 muriquis composed of various ages and sexes.[33] More often than not, smaller subgroups are sighted alone, suggesting that these muriquis may resemble those at Barreiro Rico. While there is no shortage of large trees in the Carlos Botelho forest, the undisturbed, primary vegetation is much more homogeneous than at any of the three smaller forests where muriquis have been studied. It is possible that the lower diversity of food species at Carlos Botelho limits the number of muriquis that can feed together, just as it limits the number of muriquis that the forest can support.

At each forest, food is the major force driving the muriqui's travels and grouping patterns. But, judging from the muriquis at Fazenda Montes Claros, social pressures sometimes conflict with ecological ones. At Montes Claros, the Jaó group was seen invading over 40% of the Matão group's home range during the first full year of the study, and these incursions have increased in subsequent years. Often, the two groups encountered one another at important fruit trees in the central region of the Matão group's range, and while these encounters usually resulted in both groups turning tail and departing, occasionally the desire for food was stronger than the desire to avoid a confrontation.

One especially dramatic encounter occurred at the northern limit of the Matão group's home range in 1984. The Matão group had been camped there for three days, eating Myrtaceae berries, when the Jaó group arrived. The adversaries exchanged neighs and barks when suddenly, all of the Matão males, along with just two of the females, retreated from the conflict. They traveled along a series of low ridges, then crossed the dirt road in front of the research house, and climbed up the ridge on the other side.

Three other adult females abandoned the vocal contest with the Jaó group, and began to follow after the predominantly male subgroup. I left to follow them, but neighs were still coming from the Myrtaceae ridge and I had no idea whether the Jaó group or the remaining Matão females were winning the battle. Just before these females began their descent, they stopped and looked across the valley toward the Matão males, who were neighing loudly to them, and then back up toward the Myrtaceae ridge, where the commotion was still under way. After a few moments of indecision, they reversed their tactics, returning to support the other females back on the ridge. I hesitated too, unsure of which part of the group to join, but it was late in the day and I was exhausted, so I climbed down to the road where I could hear the male subgroup from one side and the rest of the muriquis up on the Myrtaceae ridge from the other. The racket continued throughout the night, echoing around the research house as I tried unsuccessfully to sleep. At sunrise I found the males in front of the house along the road, and within 20 minutes they were back on the Myrtaceae ridge, where the entire group was reunited. The Jaó group was nowhere to be found.

In this encounter, it was clear that the interests of different members of the Matão group were at odds. The males beat a hasty retreat from the Jaó group, which had them outnumbered, while most of the females stayed to fight for their food.[34] The disparity between male and female behavior in this situation is entirely consistent with the greater importance of food to females,[35] for, although adult male and female muriquis weigh nearly the same amount,[36] gestation and lactation impose additional energy demands. When females are pregnant, they must eat enough to maintain themselves as well as their developing fetuses; once they give birth, mothers need even more energy to produce milk and carry their growing infants for months.[37] These energy requirements may explain why the diets of male and female muriquis diverge in certain months, such as October and late April, when females eat more energy-rich flowers while males eat more leaves.[38] Even among females in different reproductive conditions, behavioral differences correspond to different energy needs: Mona, the only female nursing a new infant during one year, spent significantly more time feeding than any other female; while Louise, who was neither nursing nor pregnant, spent the least time feeding and the most time traveling.[39]

Considering how important food is to females most of the time, it is not surprising that the Matão females were reluctant to surrender the sweet Myrtaceae berries to the Jaó group. The Matão males tried to lead their females away from the area, but when the females stood their ground, the males returned to their mates after spending the night by themselves.

Despite their willingness to defend food against other groups, muriquis almost never fight over it with members of their own group. Only once, during more than 1,200 hours of observing muriquis in 1983–1984, did I see anything that remotely resembled a dispute over food, and in this case, it was males rather than females which were involved.[40] One male pulled a large leafy branch away from another, but a moment later, both males began to eat from the branch together. In another instance, years later, several muriquis were eating large fruits, similar to sugar apples, that had fallen to the ground. There weren't many of these fruits around, and occasionally, while on the ground, one animal succeeded in grabbing a fruit out of the hands of another. I had never seen any behavior like this while the monkeys were in the trees, which suggests that such aggressive actions are too risky for such large-bodied animals to perform above ground.

In the canopy, where muriquis usually feed, careless movements on untested branches increase their risk of falling, and to avoid this danger, muriquis avoid feeding too close together to get in each others way. Such courteous manners are rarely seen in other group-living primates, which routintely fight for the best foods and feeding spots. Even in female-bonded species, where females cooperate in defending food from other groups, access to food within the group is determined by a strictly enforced hierarchy in which bigger males eat before females, and dominant females eat before subordinates. Muriquis resemble other female-bonded primates in their preference for—and defense of—large fruit trees, as well as in their ability to shift their diet to abundant leaves when these fruits are scarce. But, unlike these other primates, muriquis lack a pecking order; relationships between the sexes are egalitarian, and even relationships among males in the same group are remarkably peaceful. These features of their social behavior make the rules of etiquette in muriqui society unique.

6

Peaceful Patrilines

The first mating I ever observed was such a passive affair that it took me several minutes to recognize what was going on. I had been following Nancy, one of the adult female muriquis, systematically recording her approaches, first to one male, then to another, before she finally settled down on a branch less than 1 meter from one of the largest males, Irv. A moment later, Irv moved into contact with Nancy, and then nonchalantly climbed on her back. I didn't connect this unusual position with an attempted copulation until he began to thrust his hips minutes later. Nancy's son, $1\frac{1}{2}$ year old Nilo, was clearly curious about what Irv and his mother were doing, and he, too, climbed on Nancy's back to investigate. When his efforts to insert himself between Irv and his mother proved unsuccessful, Nilo swung underneath them and grabbed at Irv. Irv swatted at Nilo, who screamed and moved beside his mother on the branch. Nilo continued to scream, but Nancy's only response was to look over her shoulder at Irv, with her lips drawn back in a grin and her teeth chattering in a mating "twitter."[1] Irv resumed his sporadic thrusts, and disengaged himself a leisurely 6 minutes later (Figure 6.1).

All five of the other adult males in the group were resting in the same large tree as the mating pair, and the two which Nancy had approached before she found Irv were resting on the same branch. Both Nancy and Irv looked around while they mated, but they made no effort to conceal their liaison. The other males lounged serenely, giving no indication that they were disturbed or disgruntled that Nancy had chosen Irv instead of them as her partner.

The tolerance male muriquis display toward females and one another is the most striking feature of muriqui society. Even in what might be the most competitive of contexts, muriquis display an acceptance of one another that is not paralleled in any other primate. It is only when muriquis from different groups meet that they become hostile, with both males and females performing coordinated vocal displays and sometimes chases that emphasize their solidarity.

Within the group, however, peaceful associations are maintained as

Figure 6.1 Muriquis mate in full view of other group members.

the monkeys pursue their daily feeding and traveling activities. Affiliations are reinforced by embraces and reassuring touches, which are accompanied by friendly chuckles, but there is no evidence of dominance relationships based on threats or aggression, and no alpha male leads the group or maintains order.

One reason for the egalitarian relationships between the sexes is that male muriquis are no bigger or stronger than females. The muriqui's unusual "sexual monomorphism" means that males can't bully females into relinquishing choice food or harass them into mating. This, together with the risks of falling, inhibit aggression in all but the most critical circumstances. Ordinarily, females move freely about, avoiding competition over food by avoiding their closest associates when they feed.[2] Females also take an active role in choosing their mates by presenting themselves to preferred partners and moving away from the advances of others. Friendly males wait passively for their opportunity to mate, competing for the opportunity to father offspring in bizarrely subtle ways.

The societies of nearly all social primates are based on dominance relationships. In these societies, dominant animals supplant or threaten subordinates, particularly when they are competing for prized resources, such as places to feed and drink, and mating partners. In the ground-dwelling baboons and macaques, animals quarrel with one another even in the absence of something to fight over.[3]

Muriquis differ strikingly from this typical primate pattern. The muriquis' large bodies, so well-suited to their seasonal reliance on leaves, make it risky for them to fight in the trees. Sudden movements on to untested branches can lead to dangerous falls,[4] so it is safer for muriquis to avoid fights over food or mates simply by shifting to another spot or waiting. Although they spend 65% of their time within a 1 meter radius of one another when they are resting, they spend less than 45% in such close proximity when they are feeding.

It is rare for muriquis to act aggressively, and, even in these instances, the aggression is usually so subtle that it hardly seems like a fight. During the more than 1,200 hours of observations I logged during my first year in the forest, I witnessed only 31 interactions involving group members that could be considered even remotely aggressive.[5] Nine of these were chases, six occurring when resident females banded together to chase away subadult females from the Jaó group which were attempting to insituate themselves into the Matão group. These young immigrants typically shadowed the Matão group for several days, but whenever they tried to move closer than 10 meters, as many as five adult females would threaten them with frenzied neighs and barks and chase them off. On the one occasion when I witnessed such an incident from start to finish, the chase lasted 7 minutes and covered a distance of at least 300 meters. Males never joined in these females-only interactions, perhaps because they would welcome the addition of another mate to the group, but they didn't help the immigrants either. After a time, the resident females' efforts subsided and the company of the interlopers was tolerated.

Only three chases occurred between resident group members, and all were much briefer and covered shorter distances than the ones between residents and intruders. In one of these, at least two adult males chased a third; in another, an adult female pursued an immature; and in the third, at least three adult females joined several adult males in chasing another male. Unfortunately, however, I never detected what events had triggered these brief exertions.

The remaining 22 interactions were all of one type: one muriqui would approach another, which would move away, and the first would take the second's place. There were no obvious body postures, facial expressions, or vocalizations to indicate that these were hostile "supplants." Moreover, these episodes never involved the same pair of individuals twice. All of the supplants were initiated by either adult females (55%) or adult males (45%). Males were more likely to supplant other males (70%), while females were equally likely to supplant other adult females (25%), subadult females (25%), and adult males (25%). Adult females never supplanted another monkey in the presence of an important resource, such as food or water. In fact, the only hint that supplants might have some competitive undertones was the fact that four of the seven supplants between adult males occurred in the presence of a sexually receptive female. In three of these cases, the arriving male conducted his own inspection of the female's condition after he displaced the first male. Yet, none of the inspections that followed supplants resulted in copulations, and copulations themselves were never interrupted by supplants or any other aggressive display.

The real hallmark of the muriquis lack of belligerence is the manner in which they conduct their sexual relations. All muriqui sexual interactions are conducted openly, without any attempt to avoid being seen by other group members. In sexual interactions, like all other heterosexual interactions, females show no subordinance to males, and males are completely tolerant toward one another.

Sexual encounters begin with an inspection.[6] Males inspect the female's genital region by sniffing it or tugging at her pendulous genitalia to stimulate a liquid secretion that the male sniffs from his hand or tastes directly with his mouth. Either partner can initiate an inspection. Females often approach males and present their hind quarters, sometimes offering themselves to one male after another. Similarly, one or more males may approach and inspect a female in rapid succession.

Muriquis do not exhibit any visual cues of ovulation, such as the sexual swellings found in baboons and chimpanzees. Indeed, such signals would be of little use in an arboreal habitat where visibility is often obscured. Chemical or pheromonal signals may be the only way that female muriquis can advertise their reproductive state. At Fazenda Barreiro Rico, where males and females do not associate with one another on a daily basis, sexually receptive females engage in "urine washing," using their hands to spread their urine—and the chemical cues it carries—to branches and leaves to alert passing males to their reproductive condition.[7] But when males and females travel together, as they do at Fazenda Montes Claros, males come into contact with female scents throughout the day, and have ample opportunities to inspect females directly.

Inspections sometimes follow other friendly interactions, such as embraces. When a second male joins an embracing couple, he usually inspects the female afterward, as if he is checking to be sure he has not missed any important information that may have been transmitted prior to his arrival on the scene. Males clearly monitor—and mimic—the sexual interactions of other males, for when one male has inspected a female, another will usually try to inspect her himself.

Females are far from passive in these interactions. They have complete control over which males they interact with, and I have often seen a female avoid an inspection by one male only to present herself with suggestive grins and twitters to another.[8] Females avoid unwelcome advances simply by moving away. While males almost always oblige a female who presents herself, they do not pursue females who rebuff them, giving females a great deal of choice in determining not only which males inspect them, but also what the outcome of an inspection will be.

Not all inspections end in matings. The outcome depends both on the female's reproductive state and on the pair's mutual attraction. Unlike many other primates, muriqui males never harass females into mating with them by keeping other males away. The only aspect of muriqui sexual behavior that seems to be beyond female control is the male who inspects her but remains disinterested. In these cases, the female may crouch down with her hind end toward the male, grinning or twittering at him flirtatiously. Usually the male will respond to these entreaties, but if he is too slow, an impatient female will move on to a more ardent mate.

The parity in body size and canine tooth size between the sexes may explain why male muriquis are so complacent.[9] In most sexually dimorphic species, such as baboons and macaques, larger males can threaten both females that do not cooperate with them and smaller males that challenge the larger male's monopoly on mates. Theories about male competitive behavior, and reasons for the larger sizes of males in many species, date back to Charles Darwin.[10] His theory of sexual selection began with the observation that, in all primates, females invest far more in their offspring than males. Females bear the young and nurse them, while males, at best, may help carry the infants around or help defend the food that the females need. Rather than helping females, however, most males seem to spend their time competing with each other for females, in an effort to maximize the number of offspring they leave. Females, on the other hand, are choosier about their mates, and female choice plays a large role in determining the evolutionary fate of the species. Male aggressive behavior can be perpetuated, even accentuated, if females choose to mate with the brawniest, most aggressive male; more docile behavior can be fostered if females form strong bonds with more subtle males who protect their female friends from assaults and harassment.[11] Ironically, but not surprisingly, female choice of big, brawny males leads evolutionarily to male dominance over females. To counter this, females often group together to resist larger males, which may be one of the reasons why females in many primate species spend their entire lives among kin, relying on their matrilines for support.[12]

But in muriquis, where females are the same size as males, there is no need for females to form defensive alliances. We don't really know why muriquis, in contrast to other primates, are sexually monomorphic, but the consequences of this deviation are complex. In the first place, female muriquis can't be physically bullied, so they have more choice over who they mate with. Second, if females are partial toward males that defer to them at food sites and don't harass them unless invited, passive males will be more successful than aggressive ones.[13] Finally, if males cannot improve their chances of mating by using aggressive tactics, they must rely on more subtle forms of competition to produce the greatest number of offspring. These consequences manifest themselves in the muriqui's extraordinary sex lives.

Muriqui sex is a prolonged affair, averaging about 6 minutes, but on occasion it lasts up to 18 minutes.[14] Most of the time, the male mounts the female from the rear, but at Fazenda Montes Claros, the most sexually active female, Louise, often turned to mate face to face. In her most acrobatic act, she supported both herself and her partner from a branch with one arm, while helping him to cling to her with the other.

Whatever the position, intromission always appears to be difficult to achieve. Once a male mounts a female, both will reposition themselves until contact is made. Following intromission, long, motionless intervals are punctuated by shorter bursts of movement. The male usually grasps the female's sides, back, or shoulders, while the female looks back over her shoulder, grinning and twittering.

Not all copulations result in ejaculation, but when it occurs evidence is always visible. Males produce copious quantities of ejaculate, which clings in white, chiffon-like clumps to both the male's and the female's genitalia. The ejaculate solidifies rapidly, forming a congealed plug which, if not removed, remains visible in the female's reproductive tract for a few days, turning from white to gray to yellowish-green. Sometimes females remove these plugs with their hands immediately after mating; at other times they leave them alone. Often, both the male and female try to eat the ejaculate, perhaps as a source of protein.[15] Once, a male tried to grab a handful of his ejaculate from his partner. She screamed and turned her back on him to finish eating by herself.

A female will copulate with more than one male during the two to three day period that she is sexually active, and occasionally she will have multiple partners in quick succession. The most extraordinary demonstration of this occurred in 1990, while I was collecting fecal samples from selected females to study their ovarian hormone levels.[16] I was following Cher, a female who had given birth $2\frac{1}{2}$ years earlier, had weaned her daughter, and by my calculations should have been ovulating during the 1990 mating season.[17]

Apparently my calculations were correct. Cher was often inspected, making coy faces and twittering, and associating more closely with the adult males than was usual for her. On this particular day, I had been with Cher for nearly 6 hours, waiting for an opportunity to collect a fecal sample from her. She had defecated at least twice, but the first time she was high up in a tree and her feces splattered along the leaves and branches before reaching the ground. The second time, she was resting near some males who had also recently defecated, and I couldn't be sure which of the dung belonged to her.

At last Cher was resting on a low branch by herself. It was early afternoon, and I settled down just a few meters below her, with a clear view of her bottom. I knew she was unlikely to defecate until she was ready to move off, but I didn't want to miss such an ideal collection opportunity.

All of the males were resting in a nearby tree, and one of them, Clyde, swung over to Cher to inspect her as she lay sprawled on the branch. A moment later, he was mounting her, in one of the quickest copulations I have ever seen. Within 2 minutes it was over, and I could see the fresh ejaculate blocking Cher's reproductive tract.

Clyde swung out of the tree, and now Cutlip joined Cher. He pulled out Clyde's ejaculate, which he began to eat. Cher took a small piece of this solid material from his hand, and another dropped to the ground below. There was no longer any visible trace of a plug left in Cher. I immediately scrambled to collect the dropped piece, and when I looked up, Cutlip was already mounted and thrusting.

Within 2 minutes Cutlip had disengaged himself and swung off in the direction that Clyde had gone just a few minutes earlier. Cher had another plug in her; this time it was Cutlip's.

Next, Preta was walking along the branch toward Cher. When he reached her, he pulled out Cutlip's plug and began to eat it, with Cher taking bites while other bits dropped to the ground. Again I collected what had fallen, only to look up and find Cher mating again. When Preta moved off to follow Clyde and Cutlip, Scruff approached Cher and repeated his predecessors' performance. Within 11 minutes, four different males had copulated with Cher, and three had removed and eaten the ejaculate of the male before them. Only Scruff's plug, the last of the sequence, was still intact. As if on cue, Cher stood up, walked to the tip of the branch, and provided me with a clean fecal sample before she swung off to feed.

I was ecstatic, not only because the 6 hours I had spent following Cher were finally rewarded with a fecal sample, but because I was also able to collect pieces of the males' ejaculate plugs. I had often seen these plugs in females which had just copulated, and had even seen males and females eating pieces, but I had never had an opportunity to collect one before.

This time, I was prepared with vials and alcohol, as well as a clear view of which males had copulated and where the fragments of their plugs were dropped.

From her study at Fazenda Barreiro Rico, Milton described male muriquis waiting for their turn to copulate with one female—an astonishing behavior for any primate.[18] After nearly 30 months in the field spanning several years, this was the first time that I had seen the muriquis at Fazenda Montes Claros behaving in a similar manner. I had observed different males copulate with the same female at intervals of about 20 minutes, watching but not reacting to the copulations of one another. But I had never seen such a series of copulations—with ejaculations—in such rapid succession. Nor had I ever seen one male after another pull out and eat the plug of the male just ahead of him.

The muriquis' willingness to take turns at copulating, together with the copious quantities of their ejaculate, suggests that they have replaced overt battles over females with another kind of competition.[19] Muriquis possess extraordinarily large testes, about the size of billiard balls.[20] In other polygamous primates, there is a strong correlation between body size and testis size: the bigger the body, the bigger the testes.[21] But in muriquis, and to a lesser extent in chimpanzees, the testes are far bigger than they should be. In both of these species, overtly aggressive competition between males is constrained by the importance of cooperating. If males cannot vie with one another over access to females, then natural selection will strongly favor males who can compete in other ways. Making more sperm than the next male could be one such means of quiet competition.

Several males may copulate with the same ovulating female, but the male that produces the greatest number of viable sperm has a higher probability of fertilizing the female. Previous researchers have attributed the large testes of chimpanzees to sperm competition,[22] which may also account for the large size of muriqui testes over evolutionary time. Whether large testes let muriquis make more or better sperm is still unknown, but even without exceptional sperm, the quickly congealing ejaculate plugs physically block any subsequent suitor from easily inseminating a female. Muriquis are evidently quite capable of removing these plugs, so they are as imperfect at insuring female fidelity as most such devices made by men. In fact, depending on whether they are left in place or removed and eaten, these plugs afford females yet another opportunity to choose the fathers of their offspring.

With so much of his reproductive success riding on female favoritism, there is little a male can do to prevent another male in his group from copulating. But by cooperating with one another, males in one group can keep males from another group away from their females. And, as is the case with chimpanzees and spider monkeys, dividing mates with male kin is not as bad as dividing them with strangers, because related males share some portion of the genes that are passed on through each others' fertilizations. Consequently, male muriquis, like chimps and spider monkeys, live with their male relatives in patrilineal groups.

In muriquis, however, tolerance within the brotherhood is carried to an extreme. Unlike males in other patrilineal species, muriquis do not struggle with their brothers for dominance rank[23] and do not try to monopolize estrous females.[24] Males may have few options other than to compete for fertilizations with their sperm or ejaculate, and to monitor one another's mating behavior in order to take advantage of any reproductive opportunities a female may decide to throw their way.

One way that males can keep track of one another is by spending time together, and this is precisely what they do.[25] Since female muriquis travel in cohesive groups to search for and defend abundant food trees, males who stay with them end up together naturally. Muriquis spend more than half of their day in close proximity to at least one other individual (Figure 6.2), and maintaining varying distances from various group members helps them to avoid fights while strengthening cooperative social bonds.

The most consistent spatial relationships in a muriqui group are those among adult males, who are found within 5 meters of one another nearly 70% of the time. Subadult males, between five and seven years in age, also seek out mature males. Although adult males tolerate the approaches of subadults, their own movements are directed toward each other.

Some males are more popular than others, as indicated by the proportion of time that others are found in proximity to them.[26] But there are big differences in how males achieve their popularity. While some are initiators, routinely approaching their preferred nearest neighbors, others are more passive, rarely seeking out preferred associates themselves, but tolerating the approaches of some, but not all, other males. When the approaching male is not a friend, the passive male avoids the encounter by moving away.[27] During 1983–1984, Irv and Manga Rosa were at the top of the male "popularity hierarchy," but while Irv, a passive male, made it to the top of the list because others seemed to seek him out, Manga Rosa's popularity was a result of his own initiatives. By 1986, Cutlip had replaced Manga Rosa as the second most popular male in the group.[28] Like his predecessor, Cutlip was also an initiator.

Male spatial relations also change over time as young males mature and begin to show preferences for associating with other males.[29] Young juveniles of both sexes between $1\frac{1}{2}$ and $2\frac{1}{2}$ years of age spend over 70%

Figure 6.2 Muriquis rest in close proximity to one another.

of their time with their mothers,[30] but, even at this young age, sex differences in spatial relations exist. Juvenile males spend more time in proximity to females other than their mothers, whereas juvenile females associate more with adult males.

At some point between $2\frac{1}{2}$ and 5 years of age, males begin to shift their associations from adult females to males, and by the time they are five to seven year old subadults, their spatial relations mirror those of their adult counterparts. Juvenile females, by contrast, become increasingly more peripheral as they mature, distancing themselves from all members of the group. By the time subadult males are well on their way to establishing associations with familiar males, same aged females are prepared to leave their familiar associates and seek out a new group.

Because there are only two muriqui groups at Fazenda Montes Claros, subadult females inevitably join a group where their older female relatives have presumably immigrated. But even so, the resident adult females first repel and then ignore these young intruders. For their part, the immigrant females keep their distance for months, associating only with the subadult males and juveniles who accept them more quickly.

Although they must pass through a socially awkward migratory episode, adult females end up as sociable in their adopted groups as males are in their natal groups. Most of a mother's time, however, is spent with her young offspring, with newborn infants replacing juveniles at roughly three year intervals, but her other associations parallel those of males. Females prefer their own sex as associates, and each female has her own set of regular companions. Some females are also more popular than others, and some maintain close associations with particular males, who they may also choose as mates.

Spending time with preferred associates and avoiding others is not the only way that muriquis maintain peace. Unlike so many social primates, muriquis don't groom one another,[31] but they do offer friendly reassurances through touch.[32] They may lightly pat each other with a hand or foot when they pass in a feeding or resting tree, as if indicating that there is no need to move (Figure 6.3). Yet, the most impressive display of affection is a full-bodied embrace. In these embraces, two or more animals walk or swing toward one another and then flip upside down, so that they are hanging suspended by their tails, face to face, while they wrap their arms and legs around one another. Embraces occur in a variety of contexts: when animals move around between resting or feeding sites; prior to a sexual inspection; and during or immediately following a communal threat against an intruder.[33] At the beginning of my study, I was frequently the stimulus for these rallying displays.

Most full-bodied embraces are dyadic (84%); that is, they involve only two individuals. When mothers embrace, it is common for their semi-dependent offspring to climb around on the embracing pair. By joining in these displays, young muriquis learn their mother's preferred social associates.

There are large individual differences in the frequency that muriquis participate in embraces, and in whom they embrace, just as there are differences in their sociality and nearest neighbor associations. The majority of the dyadic embraces I observed occurred between adult females (42%), followed by those between an adult male and an adult female (25%). Nancy and Mona were active embracers, participating in over 25% of all the embracing interactions identified in 1983–1984. But, when they embraced males, Nancy was partial to Scruff (26%), whereas Mona was partial to Preta (20%).

At that time I did not know the genetic relationships between any of the adults, and I could not make sense out of many of these preferences. But in 1991, when I returned to the forest to help train two new students how to recognize the animals individually, I witnessed an embrace between an adult female and her fully grown son. I was seated with the students along a trail, waiting for the muriquis to start their day.

Figure 6.3 Handshake signifies a friendly greeting (photo by A.O. Rímoli).

The adult males were scattered throughout the tree closest to us, and I began to quiz the newcomers on who was who. They had just identified Diego, when another animal swung over to him. The students immediately saw that the new arrival was an adult female, but they needed time to study her features before they could say which one. A few minutes later, they called out correctly "Didi."

We watched as Diego and Didi reached for each other in a long, leisurely embrace. When they separated, Didi settled down on the branch beside Diego, who had turned his back to us. It was satisfying to see Diego as a robust, fully grown adult, since I still remembered when Didi carried him around as an infant in 1982. Suddenly the students looked at me in surprised understanding: this was not simply an embrace between an adult male and female; this was an embrace between a nine year old adult male and his mother.

Embraces are often, but not always, exchanged between individuals who

Figure 6.4 Six males embrace in a friendly, but animated, huddle (photo by J. Rímoli).

also maintain close spatial relationships. Nancy and Bess were one another's most frequent embrace partners, as well as most frequent nearest neighbors. By contrast, 37% of Scruff's embraces were with Nancy, yet he spent only 1% of his time in proximity to her. The importance of embraces in reinforcing social bonds is still not clear. They may be initiated by the same individual who initiates spatial proximity, perhaps to show that the approach is friendly, while at other times, they may be initiated by the individual who is approached, as a gesture of tolerance. Manga Rosa and Scruff both frequently approached other males, but Scruff participated in twice as many embraces as Manga Rosa. Most of Scruff's embraces were dyadic, with females or subadult males, while most of Manga Rosa's were polyadic, with several males in all-male huddles. Although they occur infrequently, these all-male huddles tend to escalate into intense, frenzied interactions, in which males seem to assess each others' strength with their hugs, as well as reaffirm their solidarity (Figure 6.4). At these times, the soft chuckles that accompany the calmer, dyadic embraces descend into louder, deeper throaty gurgles, or warbles.

Muriquis use vocalizations, such as chuckles and warbles, to convey both reassurance and excitement. They also call to one another to maintain contact throughout a day full of eating, traveling, and naps. Dense foliage limits vision in the canopy, and vocal communication helps the muriquis locate one another so that they can coordinate their movements. While it is difficult to know what the muriquis are saying to one another, and what they understand, they can be very noisy.[34]

Muriquis usually sleep on their haunches. Their long arms are folded tightly against their chests, their heads are turned down into their necks, and their tails are wrapped around their legs or over their heads. Several bodies are often nestled together, so that it is impossible to distinguish male from female, adult from immature, or the presence of an infant lodged snugly beside its mother. These compact sleeping clusters are divided among several trees within calling distance from one another, and as soon as they wake up, the muriquis begin to neigh.

Neighs are the most variable of the muriqui's calls.[35] Each individual has a unique set of neighs, distinct from another's and specific to each situation.[36] Other muriquis in turn react differently to neighs given by various members of the group, implying that they can recognize one another's calls. Even I could quickly pick out Sylvia's high-pitched neighs from those of the other group members.

Long, loud neighs are exchanged when the group is dispersed, while short, soft neighs are exchanged when the group is more compact. Sequences of neighs are inevitably exchanged as the muriquis wake up, as if each is announcing its presence and location. As more and more individuals become active, the frequency of these sequences increases and the neighs begin to sound more and more like a conversation. Similar discussions can be heard just before the animals settle down for an afternoon nap, and in the evening, when they have reached an area where they will spend the night.

Adult and immature muriquis of both sexes chirp softly when they are foraging and discover a particularly tasty food, and when they are feeding together in large fruiting or flowering trees. Louder chirps are sometimes emitted by the first monkey to arrive at a large patch of fruit, and I have seen other muriquis react by switching direction in mid-air and racing toward the food. Whether chirps are intentionally given to indicate the presence of food, or merely unintentional expressions of satisfaction or excitement, is not clear.[37] Whatever the intent, other muriquis alter their behavior in response to particular chirps in the same way that they alter their behavior in response to particular neighs.

There are only two situations when muriquis ignore one another's calls, and both are related to group solidarity. In the first weeks following their immigrations, subadult females lag behind when their new group starts to travel, as if they didn't understand the message to leave, and, if the group travels rapidly, they may become lost. The subadults then use long neighs to try to relocate the group, but often to no avail. It is tempting to speculate that responses are intentionally withheld by the group's unwelcoming females, but it may simply be that, because the calls are new, the group does not recognize them.

The second situation in which long neighs go unanswered is when they are coming from the Jaó muriquis far in the distance. If the muriquis are traveling, they may silently change their directions to avoid a confrontation. If they are resting when they hear the long distance neighs, they will look toward the calls, undoubtedly listening, and either remain quietly where they are, or stealthily move off.

Long neighs alert different groups to their respective whereabouts, helping them to avoid one another. Sometimes, however, confrontations are inevitable. When they do occur, they are almost always tense, proving that, although aggression is virtually absent among group members, muriquis are quite capable of hostile displays toward intruders. During the early months of my study, they repeatedly threatened me by standing upright, shaking branches, and neighing loudly. Hoarse, dog-like barks are also given in rapid, 2–3 second succession by two or more adults during a threat, and for up to 2 hours afterwards. When I first began my study, adults barked whenever I approached, and even now, if they are sleeping and I arrive silently, one sometimes gives a single, startled bark in response to my sudden appearance.

Their prolonged barking displays stopped entirely once they had become accustomed to my daily presence, but the muriquis continued to threaten visitors I brought with me into the forest. During the first year, they would clearly distinguish between me and unfamiliar humans, even when the newcomer was standing or sitting quietly at my side. Over the years, however, the muriquis have stopped distinguishing between people. They are curious when new students or visitors come to watch them, but they no longer react with vocal threats.

Relationships between the Matão and Jaó muriquis have not, by contrast, become any less bellicose over time. During intergroup encounters, members of both sexes from each group call back and forth in a haunting duet. Their neighs and barks serve to keep the two groups apart, yet, when these calls are exchanged among dispersed or nervous group members, they lead to reunions or reassuring contact. The fact that muriquis respond differently to the calls of group and nongroup members suggests that they can determine the affiliations of the callers.

Usually either one or both groups will withdraw soon after a vocal fracas begins. But when the two groups encounter one another at a large food tree or feeding area, interactions are loud, confusing, and usually militant. Adult members of both sexes participate in these encounters by neighing, barking, chasing their opponents, and touching each other. Although the contests themselves may last as little as 5 minutes, nervous barks sometimes persist on and off for two or three days if both groups remain in the same vicinity.

Over the years I have witnessed many intergroup encounters, but the most memorable ones have taken place on the ridge tops, where the low vegetation and flat terrain make it easier to see and follow the commotion. On one occasion, I was climbing up a trail to search for the monkeys when I spotted several males from the Jaó group overhead. Ordinarily, the Jaó males threatened me whenever they saw me, but this time they were silent. They moved away from me, and kept looking toward the level part of the ridge, some 200 meters ahead. Their silence was odd, and I suspected that the Matão group was somewhere nearby.

I continued toward the crest of the ridge, with the Jaó males trailing silently in the low bushes surrounding me. When I saw the Matão muriquis just ahead, and heard their soft resting neighs, I decided to stop and see what the Jaó males were up to. They were still spread out around me, and they were still completely silent. Behind me I heard a loud rustling, and suddenly one of the Jaó males was running past me on the ground toward the Matão group. He jumped into a tree, giving a sharp bark which signaled to the other males to come crashing through the underbrush. The Matão group called out in alarm, and immediately took off down the slope. I was as startled as they were by the Jaó males' sudden attack, and these "guerilla" tactics caused me to lose them for the remainder of the day.

During the last four years, confrontations between the Matão group and the Jaó group have become increasingly frequent and the outcomes more volatile. The Matão males still associate with their own females whenever they can, but six of the Jaó males have begun to challenge them with growing impunity.[38] Sometimes these challenges force the Matão males to withdraw, leaving their females in the company of their rivals until they can launch a successful counterattack days later.

Changes in the relationships between the two groups have coincided with changes in the Matão group's composition due to demographic events such as births, deaths, and migrations.[39] The muriqui's society is now in flux.

7

Life Histories, Unsolved Mysteries

One spring morning in 1990 I was watching Louise as she swung into a small legume tree next to the trail, and began eating the tiny young leaves sprouting from the branch tips. The rest of the group was scattered along the slope below her. Nilo, a young adult male, crossed into the tree to feed with Louise. He spotted some fresh shoots growing out of the ground, and descended the trunk to sit on the trail within a meter from where I stood. With a quick glance at me as I slowly sank down beside him, he began gently pulling handfuls of the tender shoots into his mouth. He was close enough to touch. I could smell his strong cinnamon scent, and hear his teeth snapping at the juicy greens. Nilo continued to feed for 5 minutes and then casually climbed back into the tree, where Louise was now resting. It had taken more than eight years for a muriqui to come to the ground so close to me.

I have known Nilo since 1982, when he was a six month old infant. I watched his mother, Nancy, wean him when he was 18 months old, and laughed at his efforts to investigate his mother's copulations soon afterward. I witnessed his first, tentative interactions with his sister, Nina, who was born when Nilo was about $2\frac{1}{2}$ years old. His younger brother, Nelson, arrived just three years after Nina, when Nilo was a $5\frac{1}{2}$ year old subadult, and by the time Nadir was born three years after Nelson, Nilo was an adult male who females sought as a mate. I have followed Nilo's development from infancy into adulthood, and while it was thrilling to have him feeding calmly at my side, it was not so surprising. He has, after all, grown up in my presence.

Nilo is one of two males and four females who were infants when my study began. Over the years, their families have grown and the group has changed. Originally, adults outnumbered immatures, but births and immigrations by adolescent females have reversed this situation. In fact, the group has nearly doubled in size, increasing from 22 to 43 members.[1]

Changes in group size and composition have profoundly affected the muriquis' behavior, just as similar demographic events such as births and deaths affect humans. Following the muriquis' life histories is like following a soap opera because, just as in humans, each individual's history is unique. For example, both Nilo and Diego, the other male infant from 1982, are still in the group, but as adults they now behave very differently. Nilo spends more of his time near his mother and the adolescents, while Diego has shifted his allegiance to the other adult males.

The reason for these differences is by no means clear. When he was five, Nilo injured his arm in a fall, and favored it for many weeks. Although he recovered fully, it is possible that this injury disrupted his normal social development, or made him more cautious than Diego in his dealings with the adult males. On the other hand, Nilo's mother may have been more attentive or protective than Diego's when they were young. I remember that, even after Nilo had been weaned and Nancy was pregnant again, Nilo was permitted to suckle until his younger sister was born, but I don't recall Diego's mother allowing this. It is also possible that the differences between Nilo and Diego have nothing to do with how they were raised, but are the result of genetics. Just as some people are timid and others outgoing, muriquis, like many other primates, exhibit distinct personalities that are difficult to explain with mechanistic analyses of their social environment.

It is the unpredictable individual differences that make nonhuman primates such intriguing subjects, but sorting out these nuances from more general patterns of behavior takes many years. Unlike cultural anthropologists, who can interview their human subjects about their personal histories, primatologists must rely on observations, which accumulate only as fast as the animals develop. And muriquis, as I have discovered, are very slow to grow up.

Muriquis remain in close contact with their mothers for at least the first year and a half of their lives.[2] Mothers are very gentle with their newborns. They use their arms to support the infant and guide it back to the nipple when it slips, because, although infants clutch at their mothers' fur during the first few weeks of life, their tails are not long enough or sufficiently developed to help them hang on.

Once they have perfected their grasping abilities, infants ride on their mother's side, close to the nipples, which are positioned almost under the arms, rather than in front as in most other primates. This anatomical adaptation may be necessary for muriquis because swinging hand over hand requires that females can fully extend and rotate their arms. It may also be safer, since females swinging enthusiastically through the canopy may crash into tree trunks and vertical branches with their bellies.

Experienced mothers are more at ease with their infants than new mothers, who seem to get thrown off balance by the unaccustomed weight. Blacky, for example, was a particularly awkward mother with her first

Figure 7.1 Fourteen month old Bernardo (photo by A.O. Rímoli).

son, Blake. During the first weeks of his life, Blake often slid down Blacky's belly, where he clung desperately to the fur of her lower abdomen. When Blacky swung into feeding trees, Blake screamed as the branches struck him, but he, like all other first born infants, has survived his mother's clumsy abuse.

By six months of age, most infants begin to ride jockey style on their mother's back. Their feet lock into place below their mother's tail, while their hands grasp the fur on her sides. Their tails wind around those of their mothers, providing extra support and balance. They are probably too big by this time to ride comfortably on their mother's belly or side without hampering her movements, and if they try to, their mothers quickly pull them around onto their back where they belong. When their mothers are resting, 6–12 month old infants are always found within a few meters' radius, awkwardly exploring their surroundings, or tucked up against their mothers for warmth or to nurse.

Between one and two years of age, young juveniles gradually exercise their independence and begin to travel short distances on their own (Figure 7.1). Mothers frequently park their offspring in the upper parts of trees, where the branches are thin enough for their tiny hands to grasp. At these times, mothers feed unencumbered, clucking reassuringly to their young by smacking their tongues lightly against the tops of their mouths between bites. Mothers also use clucks to call to their offspring when it's time to move to another tree.

Sometimes several mothers will park their young in the same place, and the youngsters play together. In fact, the offspring of females who feed and rest together may form strong social bonds with one another because of these early contacts. But not all juveniles have the same social opportunities because not all mothers are equally social. Sylvia, for example, was carrying the largest of the 1982 infants, but she did not give birth again until 1989, several years later than the other five mothers. Originally, Sylvia traveled at the front of the group, and was frequently the first animal to encounter bountiful fruit and alert the others. Since 1989, however, Sylvia spends much of her time alone with her new daughter, Sandra. Sylvia is readily accepted into the group when she decides to join her old companions, but most of the time Sandra's only company is her mother. Because of her mother's reclusiveness, Sandra's early socialization has been quite different from that of the other infants and juveniles, and following her life in the coming years will provide a natural experiment in the importance of experience to subsequent social development.

The idyllic existence between all mothers and young halts abruptly at weaning time, between $1\frac{1}{2}$ and $2\frac{1}{2}$ years of age.[3] Juveniles scream often when they are being weaned, and mothers hit or nip insistent offspring who try to suckle during resting periods or refuse to travel on their own when the group moves. Sometimes rebuffed juveniles throw tantrums, screeching and whining incessantly until either the mother relents or the juvenile tires. For weeks, their plaints fill the air.

Screeches are pitched higher than screams, and generally last longer. Juveniles screech when their mothers leave them to negotiate difficult tree-crossings alone. Small juveniles are reluctant to fling themselves across wide gaps in the canopy which larger muriquis can swing or leap across with ease. Often in these contexts, mothers respond to their offspring's screeches by using their bodies to form bridges for them to cross, and, in the early stages of weaning, young muriquis take advantage of these opportunities to latch onto their mothers. Later on, however, they run across their mothers' backs without even trying to hitch a ride.

Whines are very high in pitch and frequently last as long as 15 minutes. Group members often neigh when a juvenile has been whining too long, as if to reprimand the mother for not being more attentive. When other group members interfere vocally during a family squabble, the mothers almost always relent. Mothers are also always quick to come to the rescue if their offspring cries in distress or fear. Once, when Nilo was just two years old, he fell out of a tree right in front of me. He scurried back up the tree, screaming loudly, and Nancy lauched a violent branch-throwing display against me.

Months before they are fully weaned, juveniles begin to experiment

Figure 7.2 Juveniles play together.

with some of the foods that their mothers are eating, and by two years of age they are capable of foraging on their own. They may still try to climb on their mother's backs when the group is traveling quickly, but mothers become increasingly reluctant to let them. By three years of age, most juveniles have newborn siblings, and they are completely independent.

The traumatic transition from dependence to independence is eased by playing with other juveniles in similar dilemmas (Figure 7.2). Play accounts for only a small proportion of a juvenile's time, but it is likely to be essential to the juvenile's motor as well as social development.[4]

Most juvenile play consists of gentle grappling (90%), but sometimes it gets rougher and leads to tickling, wrestling, or chases. Grappling resembles the embraces that older muriquis engage in, and perhaps infants and juveniles pick it up when they are sandwiched between their mothers' embraces. The only difference between playful grappling and embraces is that the juveniles will also nip and tug at one another with their mouths.

When they need a respite, they may separate and hang or spin from their tails until they are ready to return to the game.

Play generally occurs during long rest periods, as independent juveniles are the last to settle down and the first to wake up. Mothers may even sleep near one another in order to provide their young offspring with access to ready play partners. In 1983, the six juveniles already in the group usually had someone their own age to play with nearby, but Molly, who was the only infant born that year, grew up with just these older—and larger—juveniles as play companions.

Although young juveniles play mostly amongst themselves (70%), older muriquis are extremely tolerant when they are solicited to play. Subadult males will sometimes join in, but adult males usually sit still while juveniles pull on their tails. When their antics have no effect, the juveniles look for more responsive partners.

Ordinarily juveniles play in pairs, but occasionally up to seven will grapple together. When they first join the group, immigrant subadult females are frequently the center of these large play groups. Soon after Cher began associating with the group, it was common for all six of the juveniles and one subadult male, Sony, to crowd around her simultaneously, tugging at her limbs and tail and bouncing off her back. Cher would remain motionless, and I suspected that she was reluctant to do anything that might attract the attention of the juveniles' mothers, who had finally stopped chasing her and were now merely ignoring her. It took weeks before Cher began to defend herself from these playful assaults, and by then she was a regular associate of Sony.

It is during subadulthood—between five and six years of age—that male and female muriquis who have grown up together generally go their separate ways. While adolescent males seek increasingly to establish bonds with the adult males in the group, adolescent females become more and more peripheral, finally leaving their mothers and familiar associates to migrate into another group.

The first females to emigrate from the Matão group did not leave until 1987, when three of the four females who were infants in 1982 reached adolescence and left. Over the years, six of the seven females born in the Matão group have left by the time they were $6\frac{1}{2}$ years old.[5] In each case, emigration occurred soon after an encounter between the Matão and Jaó groups, and three of the six females have now been observed with members of the Jaó group on several occasions. The other three are also believed to have joined the Jaó group, but their continued association remains to be confirmed.

Years before the first Matão females left for Jaó, two adolescent females from the Jaó group joined the Matão muriquis. It was late in the

1983 dry season when I first spotted an unfamiliar female, later named Cher, shadowing the Matão group a few days after the two groups had met. Each time Cher tried to approach, she was threatened and chased by the resident females, until gradually, after their juvenile offspring had begun to play with her, they tolerated her presence.[6]

Blacky followed Cher a few months later. Like Cher, she first appeared after an encounter with the Jaó group, and was threatened and chased by the resident females until the juveniles and subadult males began to play with her. Blacky was noticeably smaller than Cher when she immigrated, suggesting that she had left her natal group at a younger age. Two and a half years later, Blacky disappeared temporarily following a clash with the Jaó group, but within a few months she returned to the Matão group, where she has remained since. In 1989, Blacky gave birth to her first infant, Blake, while Cher was already nursing her second daughter. In 1991, Cher gave birth to her third infant in the Matão group.

Six other females have immigrated into the Matão group since Cher and Blacky, and each one had a similar pattern of showing up within days after a meeting between the two groups. Three of these females—Tereza, Julia, and Helena—immigrated together and have associated with each other from the start. The other three females, like Cher and Blacky, arrived separately and hung at the periphery of the group for several months. In fact, the strong parallels in the behavior of these solitary females when they first began to accompany the group has shed new light on Mona's behavior, which puzzled me during my first field season in 1982. At that time, I knew Mona had never had an infant because her nipples were not yet elongated like those of females who have nursed. She was fully adult in body size, but she usually slept in a separate tree, apart from the rest of the group, and she often lagged behind when the others started to travel. Her neighs, unlike those given by the other adult females, were rarely answered. At that time, I had no idea why Mona behaved and was treated differently from the other seven adult females present. But now, having witnessed several adolescent females as they immigrated, I suspect that Mona's odd behavior in 1982 reflected the fact that she had just joined the group.

Over time, it has become clear that, while female muriquis typically leave their natal groups, males typically remain in place—a pattern called female-biased dispersal and male philopatry. Indeed, male residency in the group has remained constant during the last decade, despite the changes in female membership. Cracking the code of the muriqui's dispersal pattern was a major breakthrough, because many features of their behavior and social organization only began to make sense once their kinship system was known.

In all primates studied to date, one or both sexes leave their natal groups when they are nearing reproductive maturity. In the majority of Old World monkeys, such as baboons and macaques, females typically remain with their mothers and sisters—forming extended matrilines—while males disperse to other groups. Most of our early knowledge about primates was based on these species,[7] and, consequently, most researchers believed that male dispersal and female-bonding were the norm. With further studies, however, the number of exceptional species has begun to exceed the rule. In monogamous species such as gibbons, and polygynous species such as gorillas and howler monkeys, both males and females generally disperse; in the multimale, polygamous chimpanzees, female dispersal and male philopatry are well documented. In fact, we now know that females migrate from their natal groups in all apes and in all New World monkeys except some capuchins, suggesting that female migration is much more common than previously thought.[8]

Despite the prevalence of female dispersal, it is still rarely found together with male philopatry. Only five other primates—chimpanzees, red colobus monkeys, squirrel monkeys, spider monkeys, and woolly monkeys—share this unusual pattern with muriquis, and the fact that the muriquis' closest relatives, spider monkeys and woolly monkeys, are among them suggests that it was practiced by their common ancestor as well.[9] Male philopatry makes sense for muriquis since muriqui brothers need help keeping other males away from independent females who cannot be bullied.[10] But explaining the benefits of male philopatry in muriquis still does not explain why females don't stay with their relatives as well.

Most anthropologists argue that animals have evolved to avoid close inbreeding, or competition with their kin for resources such as food, mates, and parental care.[11] Although this may be why females disperse when males are philopatric, it has interesting consequences for the muriquis at Fazenda Montes Claros. Here, when females move from one group to the other, they avoid mating with their fathers and brothers and competing for food with their mothers. But with only two groups in this forest, sisters end up together and granddaughters return to the same group as their uncles and grandfathers. It is possible that muriquis distinguish between close relatives and avoid mating with them,[12] and the fact that there are now adult males and their mothers in the Matão group opens up the tantalizing prospect that kinship may limit some female mate choices.

Female muriquis mate with multiple males in close succession, making it absolutely impossible to determine paternity from behavioral observations. Matrilineal kinship, by contrast, is obvious, and the genealogies that we have derived from our records of the Matão group provide provocative glimpses into muriqui family structure and reproduction (Figure 7.3).[13]

Most adolescent females are sexually active during their first year in their new group, but they do not actually begin to bear offspring until two to five years later. Because muriquis do not show any physical signs of ovulation or pregnancy, it is impossible to know whether these young females simply fail to conceive, or whether they conceive but are unable to carry their fetuses to term. Furthermore, since the exact ages of these immigrant females are unknown, it is difficult to estimate how old females are when they first reproduce.

Fortunately, however, one of the four female infants from 1982 has remained in the Matão group with her mother and younger siblings. I have no idea why this female, Bruna, has not migrated, especially since her younger sister, Brigitte, left as expected when she reached adolescence. But Bruna's odd behavior has provided a windfall. By maturing in the Matão group, we know with certainty that she was $7\frac{1}{2}$ years of age when she gave birth to her first offspring. However, we still don't know whether this is the age that most female muriquis give birth to their first infants, or whether Bruna's age at first reproduction, like her continued residence in the Matão group, is unusual.

Neither I nor the students, who were with her at the time, had any clue that Bruna had been pregnant before her infant appeared. Most adult females, including Bruna, mate during a two to three day period every four weeks. This usually only continues for two to three months, then the females become sexually inactive. Without knowing in which of these two to three day "cycles" conception actually takes place, it is difficult to calculate gestation length. Nevertheless, counting back, from observed copulations to subsequent births, gestation in muriquis should be between seven and $8\frac{1}{2}$ months,[14] which is similar to, or slightly longer than, that of the closely related spider monkeys. Longer gestation is consistent with the fact that muriquis are somewhat larger than spider monkeys, since increased body weight tends to lead to longer pregnancies across primates in general.[15]

Trying to narrow down muriqui gestation length more precisely is impossible without more specific information about female physiology. In 1990, to get a handle on this and other questions about ovulation and conception, I teamed up with Dr. Toni Ziegler, who has developed a technique to monitor hormonal cycles in female muriquis from their fecal samples. By collecting and preserving fresh fecal samples from females on a daily basis,[16] we hope to be able to identify when females are ovulating and when they are pregnant, and to correlate our hormonal findings with observed copulations and births. These data will be critical, not only to

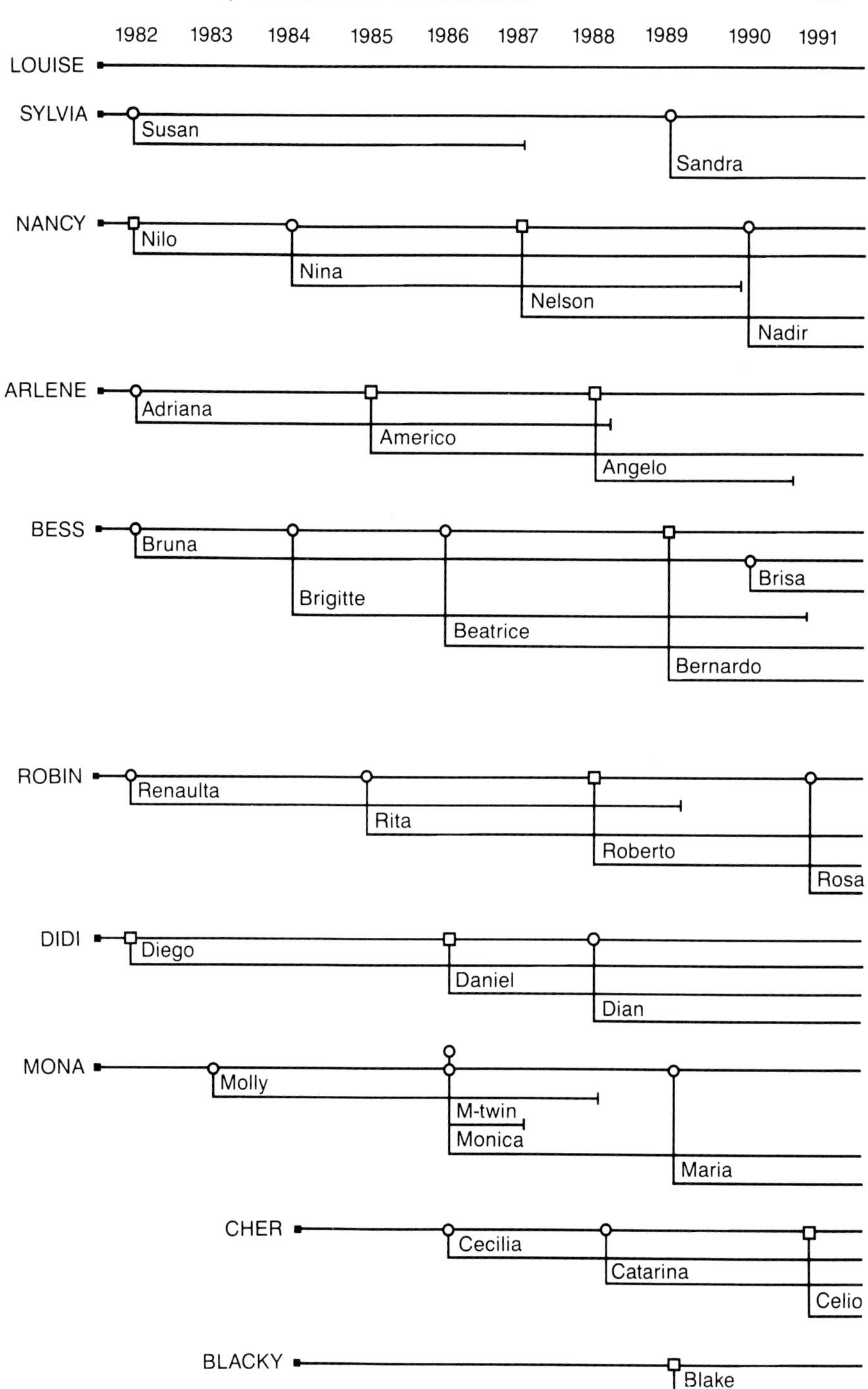

Figure 7.3 Genealogies of the Matão muriquis. Open circles indicate daughters; open squares indicate sons; terminated lines indicate disappearances or emigrations from the group.

calculating gestation length in this species, but also to understanding why different females reproduce at different rates.[17] Through detailed studies of individual female feeding preferences, we may even be able to correlate muriqui diet with reproductive success.

Seven of the eight females living in the Matão group in 1982 have had between two and four offspring in the last 10 years. Even Mona, who was in the group in 1982 but had not nursed an infant, has had three deliveries: her first in July 1983; a rare set of twins in May 1986; and a recent daughter in July 1989. Giving birth every three years is close to the average 33.8 months I have calculated for muriquis. Even Cher, who joined the group in 1983, had her first daughter in 1985, her second in 1988, and her third in 1991.

The only female who has never been seen with an infant is Louise. Oddly enough, Louise is the most sexually active female in the group, behaving as if she is cycling regularly. It is unclear why she has never had a single infant, while Bess, for example, has had four infants herself and, through Bruna, is now a grandmother.

Such differences in female reproduction may be due to any number of causes. Louise may be infertile, or perhaps unsuccessful pregnancies were never detected. By comparing her hormonal profile, obtained from the fecal samples, to those of other, more reproductively successful females, we may be able to understand what is responsible at a physiological level.

Female reproduction is related to ecology, as well as physiology, as shown by the muriquis' pronounced birth season. Of the 19 infants whose birth dates are known, 18 were born during the dry winter months of May to September. This is far more than a mere coincidence, especially since several females have had multiple births over different years, and it suggests that seasonality in the Atlantic forest strongly influences the monkeys' reproduction. During the dry winter season, muriquis feed primarily on leaves, which are low in readily available energy, but they also travel shorter distances than they do during the rainy season. It may be easier for females to give birth during the dry season, when travel demands are at their lowest. More importantly, however, females enter the most difficult period for them, when they must nurse and carry developing infants, just when higher-energy fruits and flowers become more abundant. By waiting until their offspring are 18 months old to wean them, females can rely on the following summer's bountiful fruits and flowers to replace their milk.[18]

Not all of the births have been so well timed, however. Five births occurred between my various early visits, before I had students following the group year-round, and I can only estimate these birthdates. Yet, while only one of the 19 infants whose birthdays we know was born in

Figure 7.4 Mona with her one month old twins. One disappeared at 13 months in 1987; the other is a healthy adolescent female (photo by K.B. Strier).

January—the peak of the summer rainy season—all six of the infants in June 1982 were six months old, indicating that they too had been born in January. Was the birth season different before the study began from what it has been since?

Looking back, there are other clues pointing to an unusual situation with the Matão muriquis at the very beginning of my study. For example, why were there only those six infants and two juvenile males in 1982, while now juveniles and subadults make up more than half of the group? For that matter, why has the group grown so much? In stable groups, births and immigrations are balanced by deaths and emigrations, so that the number of individuals remains the same over time. But while the eight immigrations and six emigrations have more or less balanced out, there have been many more births than deaths. In fact, in the past 10 years only five animals have disappeared without a trace from the group. We have reason to believe that two of these died; the other three remain mysteries. One of the probable deaths involved a 13 month juvenile, who was one of the twins born to Mona in 1986 (Figure 7.4). The youngster was seen just before disappearing, struggling to keep up with her mother, who was carrying the other twin on her back. Mona had managed to care for both infants through their first year of life, but she could no longer carry the two large juveniles simultaneously. Because 13 month old muriquis are still dependent on their mothers for most of their food and transportation, it is doubtful that a muriqui this young could survive on her own.

In the second case, Mark, an adult male, was seen one rainy evening in January 1988. We found him alone, more than 1 kilometer from the rest of the group, traveling with difficulty. He had a large open infection near the base of his tail. He could not reach the lesion directly with his mouth, but he repeatedly touched it and then licked his fingers. His eyes were cloudy, and at nightfall we left him resting in some bushes on the ground. We did not find him the next morning, and have never recovered his skeleton, so we can only guess about what caused his wound—and death.

The three other disappearances are more difficult to explain. Manga Rosa, an adult male, disappeared a few months after Mark, but, unlike Mark, Manga Rosa seemed healthy and was traveling normally with the group. A younger adult male, Sony, disappeared in 1991, also without any indication that something was wrong, and Angelo, a three year old juvenile, disappeared soon after Sony. None of these three males have been sighted with the Jaó group, or anywhere else in the forest despite intensive efforts to find them, but all three appeared to be in good health the last day they were seen.

Even if all five of these disappearances were, in fact, deaths, they still yield an extremely low mortality rate for the group. This low death rate, combined with their high birth rate, has resulted in a sort of population explosion in the Matão group since 1982. But what has allowed the group to grow so quickly? Or, more precisely, why was the group so small and its age structure so skewed when the study began?

It seems unlikely that hunting was responsible for reducing the muriquis' numbers. They habituated to my presence much faster than I would have expected if they had been fleeing from hunters before my arrival, and local farmers, the owner of the forest, and scientific colleagues confirmed that no muriquis had been hunted in the forest in recent years. It is far more likely that either an epidemic or severe shortage of food led to the group's depletion. Diseases have decimated the populations of other primate species at other sites,[19] and when they struck, it was the infants and juveniles that were the most vulnerable. Young primates, human and nonhuman alike, also suffer most when food is scarce. Mothers' fat reserves go down at these times, limiting their ability to ovulate, have healthy pregnancies, or nurse.[20] After a population crash caused by an epidemic or starvation, a high reproductive rate and low mortality rate would be expected among the healthy survivors, provided food availability returns to normal.

We will never know for sure whether the muriquis suffered from disease or food shortages before the study began. But we do know that their numbers are increasing at Fazenda Montes Claros, and this growth has led to three important changes in their behavior in recent years.[21] First, the muriquis have expanded their home range to include parts of the forest that were previously unused. Second, the group has begun to shift from a cohesive social organization, in which group members travel together and maintain vocal contact when they are visually separated, to a more fluid organization, in which small, heterosexual parties split off into subgroups for days at a time. Third, six adult males from the Jaó group have begun to associate with the Matão group females with increasing frequency.

Both of the first two changes make sense from an ecological perspective. Quite simply, a larger group needs to forage over a larger area because food is depleted faster when there are more mouths to feed. As long as there is suitable habitat available, a growing group will expand to exploit it. By splitting into smaller groups, the muriquis avoid competing directly at patches where food is limited. Previously, the group could stay together on a daily basis because it was still small enough to feed together in large trees, but as muriqui numbers grew, fewer and fewer trees could support them.

The group's increasing tendency to split into smaller subgroups resembles the fluid social systems characteristic of both spider monkeys and chimpanzees, as well as the social system described for the muriquis at Fazenda Barreiro Rico. It differs from these other primates, however, because, in the Matão muriquis, heterosexual associations persist even when the group has split. These associations suggest that the group will ultimately divide into two smaller, but independently cohesive, groups, each of whose social organization will resemble that of the original group during the first six years of the study. What we are witnessing right now is a group in transition.

The appearance of the six males from the Jaó group is more difficult to attribute to ecological variables. It is possible that the Matão group has become too large for the males to monitor and monopolize all the females and their offspring, especially when the females are split into smaller subgroups. The Matão males are faced with a dilemma: either they divide themselves between female subgroups or, as they have done so far, they stick together and associate with only one of the female subgroups. This strategy leaves the Jaó males free to develop liaisons with the unaccompanied females, and when the Matão females come together after a split, the Jaó males are already infiltrated. There are not, as yet, any clear patterns for predicting which of the male units will associate with which females, but the Matão males can usually repel the Jaó males when the subgroups reunite.

It is interesting that the incursions by the six Jaó males began around

Figure 7.5 Arlene before the disappearance of her son, Angelo (photo by K.B. Strier).

the time when Mark and Manga Rosa disappeared, within a few months of one another in early 1988. If the number of adult males is a variable in determining how successful a male kin unit is at preventing other males from associating with females, then the loss of these two males may have severely reduced the Matão males' ability to keep the Jaó males away.[22]

The Jaó males have brought new tensions into the group, as illustrated by an incident I observed with Dida Mendes in October 1990. Several muriquis were resting in a large fruit tree. At least eight females and immatures were there, along with the six adult males from the Jaó group. Angelo was lying along a branch beside his mother, Arlene, when suddenly one of the Jaó males approached him. We did not see any threatening gestures, but Angelo moved away and began screaming. The Jaó male followed slowly, staying a few meters behind Angelo as the youngster scrambled from one thick branch to another. Arlene made no attempt to intervene. She moved into an adjacent tree, but a new immigrant female, Fernanda, sat clucking to Angelo, who was still being pursued by the Jaó male.

The incident had a menacing quality to it. Angelo screamed continuously, clearly terrified. Finally, he managed to jump across the canopy to join his mother, and although he was already completely weaned, Arlene let him ride on her back as she swung off, down the slope.

Both Dida and I were convinced that the Jaó male was threatening Angelo, and that he could have easily quickened his pace and attacked the young juvenile if that had been his goal. We have never witnessed anything resembling this before, and the coincidental and inexplicable disappearance of Angelo only eight months later made us wonder if the events were related (Figure 7.5). We will never know for certain.

It has been both frustrating and fascinating to witness the changes in the group and the muriquis' behavior. Questions that I thought were resolved years ago are once again wide open, summoning the next decade of research. Clearly we will need more answers about muriquis if we are to succeed in fully understanding—and protecting—them.

8

Conservation Concerns and Compromises

The western part of the forest caught fire on October 4, 1990. I was with Dida and the muriquis, near the east edge of their range in a tranquil area called Raphael. I had just finished collecting my last fecal sample for the day,[1] and was writing the time, 11:11 am, in my notebook, when Jairo, the man who looks after the research house, raced up to us on the trail with the catastrophic news. Seconds later we were all running as fast as we could in our bulky field boots, binoculars crashing against our chests, down the 1 kilometer path that led back to the research house.

When we arrived at the house, Dida jumped into the car and sped off to Sto. Antonio to notify Eduardo, the director of the research station. I began filling every available container I could find with drinking water for the men who were cutting a firebreak to isolate the burning area from the rest of the forest. Moments later, a truck full of fresh recruits passed in front of the house and I hitched a ride.

We drove along the Matão road, toward the end of the forest. As we came closer, the sky grew darker from the smoke and the hundreds of hawks and vultures that were circling above, waiting for the inevitable feast of death that would soon be exposed to them. Flames leapt from the forest, which grew along the steep slopes and ridges just behind the coffee field. We turned into the field, jumped from the truck, and, grabbing as much water as we could carry, ran up the slope to where the men were clearing a firebreak about 2 meters wide. Most of them were high on the ridge trying to stem the climbing flames, while some prepared a second line of defense down on the other side. They were racing against the wind, which was blowing the fire closer and closer. Over 40 men and boys had come from the fields with their hoes and machetes to save the forest.

The fire had started in a nearby garden less than an hour earlier. Someone had set a small fire to clean out the weeds, but the wind caught it and swept it beyond the coffee field and into the forest. The coffee plants were spared because the plantation land was free of combustible undergrowth and leaf litter, but the forest was dry, and it struck like a match (Figure 8.1).

Within 30 minutes, Dida and Eduardo arrived with another carload

Figure 8.1 Fire blazed through a part of the forest at Fazenda Montes Claros, sparing the coffee plantation (photo by K.B. Strier).

of men they had picked up along the way. I knew how fast they must have driven to be back so quickly. There had not been a forest fire here in nearly 40 years, and unless the wind died down or changed direction, this one was soon going to sweep into the main part of the forest. Some of the men began setting small backfires to break up the hottest part of the blaze but it was still gaining ground.

The fire was less than 10 meters from the firebreak when it was finally brought under control. We climbed the scorched slope to bring the remaining water to the men on the other side. Backfires were burning all over, but the main fire was dying down. Most of the large trees were still standing, but they rose from the ashes of what had once been all forest. A group of capuchin monkeys had been trapped on the top of the ridge, but they had already crossed the firebreak by the time I reached the crest.

It was 2:30 pm and the immediate danger was over, but streams of smoke would continue to rise above scattered spots of simmering ashes for the next two days (Figure 8.2). A 75 acre area had burned; however, the main part of the forest was still safe, for now.

The fire was a compelling reminder of how fragile the forest—and the muriquis' future—continues to be. Even at Montes Claros, where the muriquis are protected from hunters and the forest is protected from loggers' saws, a rare accident had brought one of the last muriqui strongholds perilously close to destruction. Other muriqui populations are even more vulnerable. Those that persist in the larger state or federally

Figure 8.2 Charred remains smoldered for days after the fire was controlled (photo by K.B. Strier).

protected parks and reserves are still illegally hunted because there are too few guards to mount effective patrols. Those that have managed to survive in private forests such as the one at Montes Claros remain dependent on the goodwill of the owners and their heirs. None of these populations is truly secure.

The muriqui has become a flagship species,[2] a vivid symbol representing the entire Atlantic forest of southeastern Brazil. Muriquis appeared on the cover of the telephone directory in Minas Gerais in 1984, and on postage stamps throughout Brazil. Both regional and international media have focused increasing attention on the plight of the muriquis, leading to increased public awareness and concern about their precarious future.

Conservation groups distribute posters, T-shirts, and information to local farmers living near muriqui populations. They also lobby state and federal politicians for improved legislation to protect the muriquis. But, while education is essential, it is only one of the approaches necessary for effective conservation. Habitat protection and expansion, captive breeding, translocations, and basic research are equally critical.

Brazilian law now prohibits the clearing of any remaining stands of Atlantic forest. Some of these forests are owned and protected by state or federal agencies; others, like Montes Claros, are owned by private citizens. Publicly owned forests tend to be larger and less subject to human disturbances, but it is more difficult to protect the muriquis they support from illegal poachers. When I visited Rio Doce State Park in 1982, the guards showed me an office bulging with traps that had been removed from the forest. The Biological Reserve of Augusto Ruschi in Espírito Santo has only five guards to patrol over 11,000 acres of forest for hunters and poachers.[3] Carlos Botelho State Park in São Paulo has only 20 guards to patrol the more than 94,000 acres of forest it encompasses. The director of Carlos Botelho, Sr. Bento, lives near the park's northern border. He has been effective at eliminating hunting in this area, but in the southern part of the park it is open season on monkeys.

Apprehending illegal hunters is almost a matter of chance. In June 1990, at Carlos Botelho, three hunters were captured in the southern part of the park near the city of Sete Barros. They had entered the forest illegally on the weekend of a major soccer game, assuming that the guards posted there would be watching the televised coverage. A local restaurant that specializes in game had contracted them to supply some exotic meat. One of the hunters got lost in the forest, and fired a shot to locate his accomplices. The guards heard it, and followed it to its source. They arrested three men, but a fourth escaped.

Three muriquis were among the confiscated corpses; one was a pregnant female whose fetus was almost fully developed. The men were prosecuted and convicted.[4] But they, or others like them, will be back, and, unless additional resources are directed toward this park, it is likely that these incidences will continue to occur despite every effort by Sr. Bento and his guards to prevent them.

Ironically, private forests are better protected from illegal intruders because most of the people in the vicinity work for the landowner, usually in the surrounding plantations, and the penalty for hunting in the forest is loss of a job and a home. But some wealthy landowners have a dark side; their influence in local politics grants them immunity from prosecution if they illegally cut trees in their forests (Figure 8.3). At one private forest, several large hardwood trees were cut and removed with tractors brightly painted with the name of the mayor of the nearby city. Some of the lumber was donated for the construction of a bridge inaugurated by the mayor during his successful re-election campaign. Some of it was used to fence in one of the owner's farms, and the remainder was sold for profit. Landowners obtain tax benefits if they refrain from cutting any trees in their forests, but obviously exceptions to these restrictions can be made when they assist the very legislators empowered to enforce them.

Even the most well-intentioned landowners cannot protect their forests indefinitely. Most of the owners of the private forests are old men, who must consider their surviving families' fortunes. Sr. Feliciano, who

Figure 8.3 Illegal clearing of forest in 1984 in Minas Gerais (photo by K.B. Strier).

owns Fazenda Montes Claros, is 83 years old, and while he is still in good health, it is clear that the future of the forest will be determined by his sons. Other patriarchs have already divided continuous tracts of forest among their heirs, further increasing the number of people who must be persuaded to cooperate in protecting the land.

One solution to insure the long-term preservation of these forests is for conservation agencies to purchase the land. Most landowners and their families are understandably unwilling to sell their forests for less than their real monetary value. In 1983, Sr. Feliciano estimated his forest at Montes Claros to be worth over $1,000,000, an impossible sum for even the most successful conservation agencies to raise.

But even if the land could be purchased, it would not necessarily solve the muriquis' problems. Conservationists might eliminate selective logging, but without additional resources and a long-range plan that includes well-paid guards, hunting and poaching would undoubtedly increase once the cape of protection provided by the private landowners was removed. Legally sanctioned appropriations of private forests by the government are similarly complicated by the ill-will and loss of local protection such actions invariably provoke. It is thus not clear that the trade-off between reduced logging and increased hunting pressures is in the monkeys' best interest, especially since data now suggest that some disturbance may actually allow forests to support larger numbers of muriquis.[5]

An alternative is ecotourism, which emphasizes to the local people the benefits of preserving remaining forests and protecting the wildlife they support. Montes Claros is already a popular tourist site, and, through careful regulation, the impact of organized tour groups on the animals and our research has been minimal. Thus far, the Biological Station has been the sole beneficiary of tourist-generated income, but this income may ultimately provide Sr. Feliciano's family with further incentives to conserve their forest.

In addition to protecting existing forests that still support muriquis, there is a critical need to expand the total area of forested habitats. Remaining forests serve as natural seed banks for a variety of plant species important to the muriqui. Wind and animals can be effective seed dispersers to reforest denuded lands, but only if cleared areas surrounding the forests are also protected from plows and tractors. Purchasing these lands, or providing economical incentives to landowners to cease farming them, may be just as important as preserving the forests themselves. Management plans can also include active reforestation efforts, planting key species in areas surrounding standing forests. Reforestation can be particularly effective if it is aimed at establishing forested corridors that link isolated tracts.[6]

For a species as flexible in its diet and habitat preferences, and as adept at moving around as the muriqui, the effects of reforestation could be evident within less than one generation. In the years since my first visit to Montes Claros in 1982, I have witnessed the natural regeneration of large areas of forest that had previously been cleared. The muriquis have already expanded their activities into these areas, exploiting the new growth within them (Figure 8.4).

Apart from protecting and recovering muriqui habitat, there is concern that existing populations may be too inbred to be viable. Most surviving muriqui populations are small by biological standards, and they are isolated from one another. The smaller and more isolated populations are, the greater the risk they run of being wiped out by epidemics or by the deleterious effects of inbreeding.[7] The muriqui population in the tiny forest at Fazenda Esmeralda, for example, has declined by 33% in the last four years. Although we do not know for certain why this population is plummeting, it shows every indication of suffering from loss of genetic diversity.

If inbreeding is indeed responsible for this decline, then an obvious solution is to introduce new genes into the population by translocating muriquis from other forests to Fazenda Esmeralda. Muriquis from Fazenda Esmeralda could be exchanged for these recruits both to reciprocate gene flow and to avoid artificially overpopulating the site.

Translocations are appealing for a number of reasons, but the benefits

Figure 8.4 Muriquis at Fazenda Montes Claros would benefit from reforestation to increase the area available to them (photo by K.B. Strier).

of increasing gene flow must be carefully weighed against the risks to the individual animals and the populations involved. Capturing muriquis is risky enough,[8] but translocations may entail even greater dangers. Comparative data indicate that muriqui diets vary considerably in different forests, and translocated individuals may face difficulties in finding sufficient food and identifying edible species in an unfamiliar habitat. The pronounced differences in social organizations between different populations may hamper a translocated monkey's ability to join a group in its new forest. Whether these differences are determined by ecological variables, such as food distribution and patch size, or by local "cultural" adaptations, is still unclear, but it is critical that we at least anticipate whether or not a muriqui accustomed to living in a cohesive group can adapt to the fluid associations at another forest, or whether a muriqui from a population now lacking natural predators knows or can learn the appropriate responses to predators in another forest. There are no muriquis to waste in wrongheaded experiments, and we must know the answers to such questions before we subject healthy muriquis to such stresses.[9]

Translocations also increase the risk of spreading infectious diseases or parasites. We know from comparing different muriqui populations that both the number of parasites and the species of parasites muriquis carry vary from forest to forest.[10] But we do not know why these variations occur. Isolated populations may have evolved resistance to particular diseases that are otherwise fatal to muriquis that have had no previous exposure.

Some scientists argue that translocations are the only salvation for those muriqui populations that are already in decline. Others, however, question whether we have sufficient data to interpret natural population dynamics and fluctuations. Are efforts to save declining populations worth the risk to seemingly healthy ones? There are no sure indices for gauging how much genetic variation is necessary for a muriqui population to be viable, and the only long-term demographic data available for this species come from our work at Montes Claros. Indeed, it was at Fazenda Montes Claros that we discovered that females rather than males are the dispersing sex, so that females are the ones who should be exchanged in any translocation programs.[11]

The high birth rate and low mortality rate that have characterized the muriquis at Fazenda Montes Claros over the last decade may not be typical for this species, or even for this population. But because a reduced birth rate and high infant and juvenile mortality rate are among the first signs that a population is suffering from the effects of close inbreeding, our data suggest that inbreeding in this population has not yet had a deleterious effect. There is evidence from other primates that a population will expand when the age of females when they first reproduce equals roughly three interbirth intervals.[12] Our only insights into these important parameters in muriquis come from Montes Claros, where the eight birth intervals calculated from five different females have averaged just under three years, and age at first reproduction, known from a single female, is $7\frac{1}{2}$ years.[13] Although limited, these data suggest that the population should expand.

At Fazenda Montes Claros, the reproductive parameters indicate a viable and expanding population. Is it worth jeopardizing them, either by translocating individuals from this apparently healthy population, or by introducing foreign muriquis that may bring disease or social disruption to Montes Claros? In my opinion, it is not. The stable increase in group size and strikingly low mortality rates we have documented at Montes Claros indicate that translocations are not yet necessary for this population. If such measures are deemed necessary for the survival of the species, then I believe they should be conducted between other more imperiled populations before the Montes Claros muriquis are disturbed.[14]

Captive breeding colonies could provide another solution to preserving the muriqui. But the monkeys have been difficult to maintain in captivity, and the first facility intended for their captive breeding was established only recently at the Rio de Janeiro Primate Center.[15] It is likely that muriquis at this facility will fare better than previous captives: field studies have now provided knowledge about their diet, social organization, and reproductive behavior. Nonetheless, there are genuine concerns about where the muriquis for the captive breeding program should come from, and about what to do with the offspring that are ultimately born in captivity. The primary goal of the muriqui captive breeding project is to insure the continued existence of the species' gene pool in the event that wild populations go extinct. Ultimately, a breeding colony could supply captive-bred muriquis to replenish depleted wild populations. Such a project is not intended as a substitute for other conservation and management plans, but rather to reinforce them.

Controversy arises, however, about where to obtain the original captive breeding stock. One obvious source is the muriquis that are confiscated from people who have illegally captured them or purchased them as pets. Since 1987, four such animals have been sent to the Primate Center's facility. Three are young females, and one is a young male. All come from different regions in Minas Gerais, representing some of the genetic variation that the translocation projects hope to sustain in wild populations. The females are particularly valuable for reproductive purposes, but it took over three years in captivity before one ultimately conceived and gave birth. Some would argue that waiting for the others to mature is foolish. The current director of the Rio de Janeiro Primate Center, Dr. Adelmar Coimbra-Filho, for example, is concerned that we are wasting valuable time for the species, and he has urged that adult males should be captured from wild populations and brought to the Primate Center where they can participate in the breeding program immediately.[16]

There are some arguments to justify such action. Muriquis in certain populations, such as the one at Fazenda Esmeralda, may already be in such jeopardy that removing at least some individuals from the wild and breeding them in captivity makes sense; in their case, it is likely that they will die out if left alone. But capturing muriquis from healthy populations in more secure forests is an entirely different matter, and this difference has put some of us, who work with the muriquis at Montes Claros, into the ironic position of opposing other dedicated conservationists concerned with the future of the species. We recognize the importance of the captive breeding program for muriquis, and fully support the creation of the facilities at the Rio de Janeiro Primate Centre. Indeed, some of us were instrumental in sending one of the female muriquis to the Primate Center in 1988.[17] It is more difficult, however, to justify taking muriquis from the Montes Claros population and sending them to captivity for two reasons.

First, it seems counterintuitive to disturb the muriquis from this population when there are other populations that are far more vulnerable and where active intervention may be the only way to help them. The population at Montes Claros is growing, the monkeys still have ample forest to exploit, and the forest is well protected. If any of these conditions were to change, we would be able to document the changes, and, naturally, we would re-evaluate whether interfering with the population might benefit the muriquis.

The second, and possibly more important, reason for leaving the Montes Claros muriquis alone is their value as study subjects. Most of what we now know about muriquis in the wild comes from what my students and I have revealed. There are not yet any comparable, long-term, systematic data available from any other populations, and it seems counterproductive to jeopardize this essential source of information as long as there are other populations that can be manipulated without disrupting ongoing research.

Throughout our research, we have avoided any manipulation of the animals, even though this has made our task of studying them more difficult. It would have been quicker to identify and age individuals, for example, if we had trapped and marked them with tattoos or collars and taken dental casts. But, by waiting until we could recognize individuals by their natural markings and document maturation in muriquis of known age, we avoided subjecting them to the risks of live capture. Similarly, there would have been fewer days spent searching for the group if we had put radio transmitters on the monkeys, but instead we waited until we learned their travel routes and how to follow them.[18]

While a species as endangered as the muriqui will ultimately need active management, no plan can succeed without detailed knowledge of the monkeys and their ecosystem—knowledge that only long-term field studies can provide. I was introduced to muriquis by conservationists Russell Mittermeier and Celio Valle,[19] who recognized that data from such a study of the muriqui would contribute significantly to the development of effective management plans on their behalf. Both the study at Montes Claros, and the few comparative studies that now exist, have already provided vital insights into muriqui ecology and behavior that will guide our conservation efforts. At present, we are still limited in our ability to generalize from the Montes Claros population to the species at large, and both the continuation of the research at this forest and the initiation and continuation of comparable studies of other populations will be essential to the muriquis' survival. Indeed, the ways in which the research can contribute to their conservation has become an increasingly strong motivation to continue the project at Montes Claros.

The long-term scientific presence we have sustained at Montes Claros has not only provided information critical to the conservation of muriquis, it has also protected them and their Atlantic forest indirectly. Our presence demonstrates the importance of this forest and the muriquis it supports to local Brazilians and government officials.[20] At the same time, we have respected the needs of local people by not interfering with the small-scale selective logging that has continued over the years, and, because of these respectful relationships, we have been able to monitor its occurrence and its effects on the muriquis. Our research has also insured that tourists who come to Montes Claros can see the monkeys. Without us, no one would know where to find the monkeys, and the muriquis would be far less tolerant of unfamiliar visitors. Ecotourism has increased dramatically in recent years, sometimes threatening to compromise our research objectives. We have had to restrict tourists to the Matão road and to a path leading into the Raphael portion of the forest, in order to give the muriquis the option of moving away from the visitors, into the forest, if they are disturbed. Yet, despite the disruption tourists bring, their financial contributions to the research station provide income essential to its maintenance. And tourists who see the muriquis and other wildlife in this rare ecosystem take away with them a deeper appreciation of nature, which we hope will translate into stronger conservation policies.

In addition to these indirect benefits, the scientific research at Fazenda Montes Claros has provided insights into muriqui ecology and behavior that shape any conservation plans. For example, one of the original goals of the research was to characterize the muriquis' diet fully, because the proportion of different food types in their annual diet, the spatial and temporal distribution of important food species, and the ways in which muriquis exploit their sources of food all have important implications for the development of management plans as well as for understanding their behavioral ecology. While it may not be feasible to provide captive muriquis with the same food species they exploit in the forest, efforts can be made to approximate the protein, carbohydrate, fat, and fiber composition of their diets.

Identifying important food species is also critical to establishing new muriqui populations in protected forests. Existing forests that do not now support muriquis may, in fact, be unable to sustain transplanted populations or individuals that are reintroduced from captivity to the wild. We know from the few comparative studies that have been conducted so far that muriquis feed on a variety of species, and they occupy habitats with considerably diverse plant communities.[21] In fact, contrary to popular belief, the forests best able to support the largest muriqui populations may not be those that are the biggest or most pristine. Primate home range sizes are related both to population sizes and to the distribution and density of edible foods, and for muriquis it is difficult to separate out these influences and know when a population is close to overexploiting its available food. Small forests with high-density food resources may be able to support larger populations of muriquis than larger forests with more sparse resources, and these smaller forests may ultimately prove more useful for preserving muriqui numbers.

Nevertheless, growing populations can certainly become too large for the available food supply, particularly in isolated islands of forest where there is no room for expansion. When animals compete over limited food, mortality may rise and fertility may fall, with dire consequences for the population's ability to recover. While there is still abundant, productive forest available to the muriquis at Montes Claros, we must continue to monitor the population closely for any evidence that it is at risk of catastrophic collapse.

One way to prevent such disasters is to increase the size of the forest available to these and to other muriquis through reforestation of surrounding areas, but to make these reforested areas useful to muriquis requires an understanding of their dietary preferences. Another way is to link together forests that can support muriquis so that the animals can expand their ranges, but this requires knowledge of muriqui relationships within their ecological communities. In some forests, for example, muriquis may be important prey for large predators, such as jaguars and ocelots. Because these predators require large hunting ranges, they are usually most vulnerable to forest fragmentation, and they disappear sooner from disturbed forests than do primates such as muriquis. In areas where predators have been eliminated because of forest fragmentation, or hunting, muriqui populations may be released from the weight of predation, and their numbers are likely to be more sensitive to the availability of food and to disease. Yet, while the disappearance of predators may initially benefit muriquis, it is not clear over the long term whether their populations are healthier, or at greater risk, than those culled by predators.[22]

In a similar way, primates have evolutionary relationships with the parasites that live in their gastrointestinal tracts. Initially, I was relieved to discover that the muriquis at Montes Claros were free of intestinal parasites, and surprised that muriquis at the larger, less-disturbed forest at Carlos Botelho had a higher prevalence of infection. But, in fact, these observations may indicate that the complex relationships between parasites and their muriqui hosts have been disturbed at Montes Claros, and how this may affect the monkeys is not yet known.[23]

Muriquis also have complex relationships with other primates. We know from our work at Montes Claros that muriquis exploit many of the same foods as the other monkeys found here. This knowledge is useful for identifying potential forests where muriquis can be transplanted or reintroduced. A forest that supports brown howler monkeys, for instance, is likely to be a suitable muriqui habitat, because the diets of howlers and muriquis overlap substantially. But even though muriquis introduced to such forests may survive, they may compete with resident primates for food, and ultimately cause declines in these populations. Brown howler monkeys, like many other Atlantic forest animals, are also highly endangered and in need of protection, and it is not immediately obvious that muriquis should be protected at the expense of other species.[24]

Ecological differences, such as the spatial and temporal distribution of food, can explain at least some of the variation in the size, composition, and cohesion of muriqui groups between sites,[25] and the shift toward more fluid grouping patterns at Montes Claros can be attributed to the increase in group size.[26] Understanding the variations in grouping patterns and social organizations that occur in wild populations is important to understanding what constitutes a functioning social and breeding unit in the muriqui. Past experiences with other primates brought into captive breeding programs without prior knowledge of their natural social and mating systems have had mixed success. In a species like the muriqui, which matures slowly, knowing their natural reproductive patterns is critical to distinguishing viable wild populations from those that are doomed without active intervention.

Yet, we still don't really understand the bases for muriqui social organization. Despite observed differences in grouping patterns, muriquis at all sites where they have been studied exhibit strikingly low levels of intragroup aggression. The low levels of aggression may be related to kinship among males and mutual avoidance of competitive situations among females.[27] It is not clear, however, how tolerant unrelated males will be if confined together in captivity. Similarly, the coincidental occurrence of the disappearances of the two adult males in 1988 and of changes in the social organization of the Montes Claros muriquis suggest that the size of male kin groups may be critical to the ability of males to monopolize females for reproduction.[28] These observations provide indirect evidence that the removal of even a single male might artificially alter the competitive abilities of his remaining relatives, stressing an otherwise stable group of males.

Understanding muriqui kinship and dispersal patterns, like most of the other results that are described in this book, has taken many years of patient study. These findings no doubt will contribute to the success of conservation plans on behalf of the muriqui. But, even as our research progresses, it is astonishing that there is still only one ongoing long-term field study of a single population, which may or may not be representative of the species. Perhaps the most urgently needed conservation projects are other long-term studies so that we can compare groups.

But conservation is costly in both time and money, and we may no longer have the leisure to concentrate exclusively on more long-term studies. Indeed, many conservationists acknowledge the importance of such research, but are simultaneously anxious to initiate more active management programs with more immediate—and sensational—results.

All conservationsists agree that efforts must proceed simultaneously at multiple levels, including breeding muriquis in captivity, managing existing populations, protecting and expanding habitats, and conducting further research. But there is strong disagreement about which should take priority, and, particularly, about where to focus limited financial resources. Brazil, like most of the other countries that are home to endangered primates, suffers from overwhelming economic difficulties. As a result, conservation relies on the governments of wealthier countries, and on private organizations that can raise the necessary funds. Such private agencies, which have funded the majority of conservation projects on behalf of the muriqui, must spark the interest of donors to raise money, and high-profile plans of action generally attract more attention than basic field research, which requires an investment of many years before interesting findings begin to emerge.

Within the last decade, private industry and international agencies have assumed an increasing role in conservation. Environmental impact surveys are now necessary before any internationally funded development project can begin in Brazil, and many private Brazilian companies have sponsored similar studies independent of external pressure.[29] While it is not clear how effective such surveys will ultimately be in protecting habitats and wildlife, they are nonetheless an appropriately more cautious move toward informed development.

Perhaps the biggest obstacle to conservation in Brazil is the country's exorbitant international debt. Recent innovative approaches in a number of countries have linked reductions in foreign debt to conservation commitments, in what are now called "debt-for-nature" swaps. While the terms of these agreements are tailored to each case, the general format is simple: International conservation agencies purchase some of a country's foreign debt by paying a fraction of its real value to the banks to which it is owed; the country then repays the conservation agencies by making a proportionate contribution, in dollar value, to conservation.[30] Such exchanges have already been implemented in Bolivia, Costa Rica, and Venezuela, but until recently the Brazilian government was perceived as being too unstable. When a debt-for-nature swap was proposed in 1988, the then President of Brazil, José Sarney, stated unequivocally that he was not interested in opening up opportunities for foreign control in Brazilian territory.[31] With a change in presidency, a rewriting of the Brazilian constitution, which now contains a full chapter on environmental protection, and increasing concern among Brazilians about the preservation of their natural resources, debt-for-nature negotiations have resumed.

International attention has focused on the Brazilian Amazon, the largest remaining tropical forest in the world, in part because of the consequences that deforestation of such an extensive area may have on global climate. Compared to the Amazon, the Atlantic forest, and the muriquis that both represent it and depend upon it, continue to receive relatively little attention. Yet, what has happened to the Atlantic forest and the muriquis that inhabit its tattered remnants is a far-reaching reminder of how quickly a natural heritage can be destroyed. It remains to be seen how quickly, and how well, we can learn about the muriquis and their endangered ecosystem so that we can protect them.

After a decade of research, our work with the muriquis has just begun.

Appendix

FORESTS KNOWN TO SUPPORT MURIQUIS

Forest (numbered location on map*)[a]	Status	Acres	Muriquis
Fazenda Esmeralda (5)[b]	private	100	12
Fazenda Córrego de Areia (1)[c]	private	150–337	8
Simonésia (6)[d]	private	2,000	21+
Fazenda Montes Claros (3)[e]	private	2,000	80+
Cunha State Reserve (8)[f]	public	5,575	16+
Fazenda Barreiro Rico (9)[g]	private	8,147	95+
Augusto Ruschi Biological Reserve (4)[h]	public	11,125	10+
Caparaó National Park (7)[i]	public	43,670	12+
Jureia Ecological Station (11)[j]	public	50,000	4+
Rio Doce State Forest Park (2)[k]	public	87,500	21+
Carlos Botelho State Park (10)[l]	public	94,110	132+

*See Figure 2.2, and Chapter 2.

[a] Populations are from Mittermeier et al. (1987), except where indicated by more recent data.

[b] See text for decline in population from high of 18 in 1987–1988 to a low of 12 in 1990.

[c] See text.

[d] See text.

[e] The population has increased since Mittermeier et al. (1987), when it was estimated at 52+.

[f] This reserve was established in 1977 under the administration of the State Forestry Institute. A total of 11 muriquis have been sighted at this site (Mittermeier et al., 1987).

[g] See Torres de Assumpção (1981, 1983a, b), Torres de Assumpção et al. (1982), Milton and de Lucca (1984), and Milton (1984b).

[h] The Reserva Biologica Augusto Ruschi is administered by the Brazilian Institute for the Environment and Renewable Natural Resrouces. It is located at 19° 38′–46′ S, 40° 40′–45′ E, near the city of Santa Teresa. It supports five species of primates, although according to a recent 12-month census (Pinto et al., submitted), muriquis occur at lower densities than the other species.

[i]A group of 12 muriquis was located just outside the park's boundaries, but none have yet been sighted within the park. Mittermeier et al. (1987) suggest that Caparaó is an ideal site to develop ecotourism because many visitors already come to the park to climb the Pico da Bandeira, which, at 2,890 meters, is the highest peak in southern Brazil.

[j]Jureia is one of the forested areas in São Paulo administered by the state's Instituto Florestal under the Secretary for the Environment of the Federal Government. Only 4 muriquis have been reported, but the site and remoteness of this forest strongly suggest a larger population (Mittermeier et al., 1987).

[k]This park is administered by the State Forestry Institute. It was created in 1944. See Stallings (1988) for more details about the area and the small mammal community.

[l]This park was established in 1982 under the administration of the State Forestry Institute. Four smaller areas that had been protected reserves since 1941 were fused to create the park. See Paccagnella (1986).

Notes

NOTES TO CHAPTER 1

1. Translated from Camara Cascudo (1985:60): *Toda a gente se admira/Do macaco andar em pe/O macaco já foi gente/Pode andar como quiser!* The superstition that monkeys were once human is widespread throughout Brazil and Africa. There are several Brazilian variations of this motif; none appear to refer exclusively to the muriqui.

2. Some recent weights from muriquis captured and then released are substantially lower; see Lemos de Sá and Glander (in press).

3. Cant (1977, 1986) was the first to propose the trade-off between energy expenditure and time minimization for spider monkey brachiation. Also see Parsons and Taylor (1977) for energy costs.

4. See Hill (1962) and Zingeser (1973).

5. See Bauchop (1978), McNab (1978), and Parra (1978) for discussions of leaf processing in primates; also see Richard (1985) for more general discussion.

6. For discussion of the relationships between body size, metabolic rates, and diet, see Gaulin (1979) and Western (1979). For general discussion of primate diets, see Richard (1985).

7. See Hill (1962), Zingeser (1973), Milton (1985a), Kay et al. (1988), Rosenberger and Strier (1989), and Lemos de Sá and Glander (in press).

8. See Trivers (1972), Clutton-Brock et al. (1977), and Leutenegger and Kelly (1977). Also reviewed in Gray (1985).

9. See Strier (1990, 1992).

10. See Wied-Neuwied (1958), and Aguirre (1971) for various regional names and spellings for the muriqui. In the earlier accounts, the muriqui was sometimes assumed to be a member of the genus *Ateles*, probably because of its close physical resemblance to the spider monkeys (see also Moynihan, 1967).

11. For the most recent published description of remaining muriqui populations, see Mittermeier et al. (1987). These forests are briefly described in Chapter 2 and the Appendix.

12. See Rosenberger and Strier (1989), and Strier (in press, a) for reviews of the relationships between the Atelinae.

13. See Wolfheim (1983) for descriptions of the geographical distributions of the Atelinae.

14. The other Primate division is the Strepsirhines. Living members include lemurs, bushbabies, and lorises, all of which are now restricted to Africa and Asia. Tarsiers are usually included with the Haplorhines rather than the Strepsirhines. For an excellent review and bibliography of primate evolution and taxonomy, see Fleagle (1989), and Conroy (1990).

15. See Ciochon and Chiarelli (1980) for a thorough discussion of the alternative perspectives on the origin of the New World monkeys, or Platyrrhini.

16. See Camara Cascudo (1985).

17. Recent DNA hybridization and sequencing techniques have revealed that humans share roughly 98% of their DNA with chimpanzees and gorillas. Which of these apes is more closely related to humans remains controversial. For further insights into these controversies, see Sibley and Ahlquist (1984), Lewin (1984, 1988a, b), Miyamoto et al. (1987), and Marks (1991).

18. See Eisenberg et al. (1972), Clutton-Brock (1977), and Clutton-Brock and Harvey (1977, 1979, 1984) for examples and discussions of comparative analyses in primates.

19. The status of animal populations is assessed through census data and reported sightings. The International Union for the Conservation of Nature and Nature Resources (IUCN) classifies species as: *extinct* if they have not been located in the wild within the past 50 years; *endangered* if numbers or habitats are reduced to critical levels and the species is in danger of extinction; *vulnerable* if populations or habitats are decreasing and will become endangered if the causal factors are not stopped; and *rare* if they are at risk because of restricted geographical or habitat distributions. See IUCN (1982), and Mittermeier et al. (1982).

NOTES TO CHAPTER 2

1. For more details about deforestation in the Atlantic forest region, see Mori et al. (1981), Coimbra-Filho (1984), and Fonseca (1983, 1985).

2. All of the information presented in this section is summarized from Hatton et al. (1983), Fonseca (1983, 1985), and especially from Rizzini and Coimbra-Filho (1988). See also Mori et al. (1981).

3. The theory of forest refuges was originally proposed on the basis of the biogeography and speciation of South American birds (Haffer, 1969). See Kricher (1989:151–153) for a concise review.

4. See Haffer (1974).

5. See Müller (1973), and Jackson (1978).

6. See Mori et al. (1981).

7. See Kinzey (1982).

8. See Kinzey (1982) for a detailed review of Atlantic forest refuges and their effects on primate endemism.

9. *Brachyteles* is one of two primate genera endemic to the Atlantic forest. The other is *Leontopithecus*, or the lion marmoset, which has three distinct subspecies that probably deserve separate species status (Rosenberger, 1981; Emmons, 1990). The masked titi monkey, *Callicebus personatus*, and the brown howler monkey, *Alouatta fusca*, are the other two endemic primate species.

10. Vieira (1944) originally suggested two subspecies for muriquis, but later he (Vieira, 1955), and others (e.g. Hill, 1962; Napier, 1976; Kinzey, 1982) concluded that there is insufficient justification for two subspecies.

11. For thorough reviews of human impact in the Atlantic forest, see Mori et al. (1981) and Fonseca (1983, 1985). Most of this section is summarized from Fonseca (1983, 1985).

12. See Wied-Neuwied (1958:349).

13. See Wied-Neuwied (1958:78).

14. See Wied-Neuwied (1958:78).

15. See Wied-Neuwied (1958:108).
16. See Wied-Neuwied (1958:388).
17. See Fonseca (1983:35).
18. See Wied-Neuwied (1958:388).
19. See Aguirre (1971) for a rich description of muriqui history and regional lore, as well as for the results of the first extensive census of remaining populations.
20. See Coimbra-Filho (1972).
21. See Mittermeier et al. (1987).
22. Two additional muriqui populations have been reported in state-protected forests in São Paulo. One of these is contiguous with Carlos Botelho State Park.
23. See Lemos de Sá (1988).
24. Lemos de Sá (personal communication).
25. The Brazilian Foundation for the Conservation of Nature (FBCN) recently "loaned" this area to Fundação Biodiversitas, which maintains an active administration and involvement with the Abdalla family, and the research.
26. See Strier (1991a); also discussion in Chapter 7.
27. See Chapters 5 and 8 for more detailed discussions.

NOTES TO CHAPTER 3

1. See Wrangham (1979, 1980), and van Schaik and van Hooff (1983).
2. Models for ungulates were developed by Jarman (1974); for birds by Emlen and Oring (1977); and for bats by Bradbury and Vehrencamp (1977).
3. See Dittus (1977), Mori (1979), and Small (1981).
4. Mittermeier is currently the president of Conservation International; in 1981, Mittermeier was responsible for the World Wildlife Fund film, *The Cry of the Muriqui*, made by Andy Young, to publicize the Brazilian Atlantic forest.
5. To conduct research in Brazil as a foreigner it is necessary to have Brazilian sponsors. Professor Celio Valle sponsored my dissertation research period; Dr. Coimbra-Filho and Admiral Ibsen also provided essential assistance. Permission from the National Research Council (CNPq) is required before foreigners can conduct research in Brazil. It usually takes six to nine months for a project to be approved.
6. See Valle et al. (1984), Alves (1986), and Santos et al. (1987) for examples.
7. See Hatton et al. (1983) for a thorough description of the vegetation at Montes Claros.
8. See S.L. Mendes (1985, 1989) for research on the brown howler monkeys, and Ferrari (1988) for research on the buffy-headed marmosets.
9. See Nishimura (1979).
10. In addition to Mittermeier, film-maker Andy Young was along to take photographs and collect preliminary data for his Bachelor's thesis, as were Dr. Devra Kleiman of the National Zoo, Jon Jensen, then of the Jersey Wildlife and Preservation Trust, and Mittermeier's Brazilian assistant, Carlos Alberto M. Pinto. Mittermeier, Kleiman, and Jensen left Brazil after a few weeks; Andy and I remained with Carlos Alberto for the following two months.
11. Capuchin monkeys are called *macaco pregos*, or monkey nails, in this region because of the shape of the penis during the displays males give when they are in an aroused or alarmed state. Howler monkeys are called *barbados*, or beards, because of the long hair covering their large throat sacs.

12. Carlos Alberto Machado Pinto is an ex-hunter who was a university employee working for Celio Valle and his group. He routinely accompanied Mittermeier throughout Brazil until his recent retirement.

13. See Clutton-Brock and Harvey (1977, 1979) for some correlates of folivory in primates.

14. See Hill (1962), and Zingeser (1973); also Chapter 1.

15. See Altmann (1974) for classic discussion of scan sampling methods.

16. Transforming the trail measurements into a trail map was a time-consuming process. The distance between each measurement had to be reduced in proportion to the scale drawing, taking into account the inclination of the slope, and the compass direction had to be measured with a protractor.

17. See Altmann (1974) for discussion of focal animal sampling. I could not conduct continuous focal sampling because of the logistical difficulties of keeping the animal in view and recording its ongoing behavior. For this reason I recorded data on-the-minute for focal animals.

18. See Terborgh (1985) for similar, but independently developed method; see Strier (1989) for details on the Feeding Tree Focal Samples (FTFS), a method I developed with the help of Richard Summers.

19. See Strier (1986, 1991b) for detailed description of phenological data collection.

20. Fisher and Yates (1957).

21. See Leighton and Leighton (1982), Symington (1988), White and Wrangham (1988), and Strier (1989).

22. A regression analysis indicated that canopy volume can be reliably predicted by DBH from the equation: $y = 1.97x - 0.67$, where y = log canopy volume and x = log DBH; see Strier (1989).

NOTES TO CHAPTER 4

1. Funding for my dissertation research was provided by the National Science Foundation, the Fulbright Foundation, the Joseph Henry Fund of the National Academy of Sciences, Sigma Xi, and the World Wildlife Fund.

2. See Chapter 2.

3. Rosa Lemos de Sá went on to conduct her Master's research on another muriqui population in Minas Gerais. See Lemos de Sá (1988) and Lemos de Sá and Strier (in press).

4. See Fonseca (1983) and S.L. Mendes (1985, 1989).

5. Nadir Eduardo Ferreira now lives in Caratinga. Her younger sister, Lada, has replaced her at the research station.

6. Jairo Gomes continues to work at the research station, and recently he has begun to participate in the muriqui research.

7. I adopted the same system of naming that was employed on the Amboseli baboon project in Kenya, which I had worked on during January through June 1979.

NOTES TO CHAPTER 5

1. Muriquis usually sleep late into the morning on cold winter days, but occasionally they wake up and move away from their sleeping area early, and the

only way to be sure of not losing them on these rare occasions is to be there, waiting, just after dawn. During the middle of the winter in 1988, some 25 well-known primatologists who were in Brazil for the International Primatological Society meetings came to visit me and the muriquis at Montes Claro. I took small groups of visitors out for a few hours each with the muriquis. It was a fascinating experience, because each of these primatologists brought new perspectives: Chuck Snowdon described the muriquis' behavior in terms of their vocal communication; Frans de Waal interpreted their interactions in terms of reconciliations (see de Waal, 1989); Toni Ziegler began thinking about their reproductive physiology; Richard Wrangham compared them to chimpanzees; and Stuart and Jeanne Altmann compared them to baboons. It was hard on some of the primatologists who were not accustomed to fieldwork. One who often came out at 6:00 am and had a five hour wait before the monkeys became active, fell soundly asleep beneath the muriquis.

2. See Strier (1987a).

3. See data in Strier (1991b).

4. For discussion of the effects of seasonality on body size, see Lindstedt and Boyce (1985).

5. See Torres de Assumpção (1981), and Ferrari and Strier (1992).

6. See various chapters in Chivers et al. (1984).

7. See Strier (1989).

8. Two important legume species that muriquis exploit heavily for new leaves contained no detectable levels of tannins in analyses performed by Dr. Eloy Rodriguez (see below).

9. See Milton (1984a) for muriqui food passage rates.

10. Two detailed studies spanning more than one annual cycle regarding the muriquis' role in seed dispersal have recently been completed by José Rímoli at Montes Claros, and Pedro Luis Rodriguez de Morães at Carlos Botelho.

11. See Janzen (1966), and Kricher (1989).

12. Ethnobotany is a historical discipline with numerous accounts of various quality and specificity. For a recent and highly systematic example, see Schultes and Raffauf (1990).

13. See Phillips-Conroy (1986) for medicinal plant use by baboons; and Wrangham and Nishida (1983), and Huffman and Seifu (1989) for medicinal plant use by chimpanzees.

14. See Strier and Stuart (1992), and Stuart et al. (in press).

15. Dr. Eloy Rodriguez of the University of California-Irvine is currently examining select muriqui plant foods for bioactivity. See session, organized by Rodriguez and Wrangham on Zoopharmacognosy: Medicinal Plant Use By Wild Apes And Monkeys at the 1992 annual meeting of the American Association for the Advancement of Science, and described by Gibbons (1992).

16. See Clutton-Brock (1977).

17. See Strier (1987b).

18. For discussion of energy costs of suspensory locomotion, see Parsons and Taylor (1977); also see Chapter 3.

19. See Cant (1977, 1986), Milton (1980), and Chapter 3.

20. See S.L. Mendes (1989).

21. See Strier (in press, a).

22. See Hatton et al. (1983), and Chapter 3 for description of habitat types; and Strier (1987b) for data regarding the muriquis' use of different habitat types.

23. See references in Chapter 2.

24. For example, muriquis at Montes Claros occur at a density of 0.1 individual/hectare. In the tiny, highly disturbed forest at Fazenda Esmeralda, the muriqui density is 0.3 individual/hectare. In the large, primary forest of Carlos Botelho state park by contrast, muriquis occur at a density of only 0.03 individual/hectare.

25. See Milton (1984b).

26. Eduardo Veado and Cristina Alves accompanied me on this trip. For further descriptions of the study site, see Torres de Assumpção et al. (1982).

27. See National Academy Press (1981).

28. For comparison of DBH sizes at Montes Claros and Barreiro Rico, see Strier (1986).

29. See Lemos de Sá (1988).

30. See Lemos de Sá and Strier (in press).

31. See references in Chapter 2.

32. Luiz Paulo S. Pinto and Claudia M.R. Costa were the two students who spent a year looking for muriquis at Augusto Ruschi. See Pinto et al. (submitted) for the results of their survey.

33. Sandra G. Paccagnella, Pedro Luis R. de Morães, and Oswaldo Carvalho Jr. are the students who have worked at Carlos Botelho since 1988. See Paccagnella (1986) for an earlier census.

34. The two females, Mona and Cher, which stayed with the males may have had specific motives for doing so. Mona was the only female nursing a young infant this year, and she may have been trying to prevent any harm from befalling her daughter during the ensuing dispute, despite the fact that she would lose the chance to continue feeding on the Myrtaceae ridge in the process. Cher was the only recent immigrant in the Matão group at this time, and may have followed the males because she was still not tolerated by the other females in her new group.

35. See references in Chapter 3.

36. See references in Chapter 1; and especially Rosenberger and Strier (1989), and Lemos de Sá and Glander (in press).

37. See Altmann (1980), and Demment (1983).

38. See Strier (1991b) for sex differences in diet. It is also possible that phytoestrogens are important in regulating female fertility. See reference to Strier in Gibbons (1992).

39. See Strier (1987a).

40. See Chapter 6.

NOTES TO CHAPTER 6

1. See Milton (1985b).
2. See Strier (1990).
3. See review in Gray (1985), and de Waal (1986).
4. See Strier (1992); Chapter 5.
5. See Strier (1992).
6. See Strier (1987c).
7. See Milton (1985c).
8. Strier (1992).

9. See Milton (1985a).
10. See Darwin (1871), and review in Gray (1985:193–272).
11. See Smuts (1985, 1987).
12. See Strier (1990).
13. See Strier (1992).
14. See Strier (1987c).
15. Drs. Deanne Mosher and Bill Checovich, of the University of Wisconsin-Madison Medical School, are currently analyzing the ejaculate samples I obtained from the muriquis in 1990. See Milton (1985b).
16. In an effort to understand the physiological bases of muriqui reproduction, Dr. Toni Ziegler and I are examining the ovarian steroids present in their feces. Fecal assays provide a noninvasive way of determining whether females are ovulating, pregnant, or lactating. See Chapter 7 for methods of fecal collection.
17. Based on their known reproductive histories, I suspected that Cher, Robin, Didi, and Arlene would be ovulating during the 1990 mating season. In fact, all four females were sexually inspected by males during this period, and so far, both Cher and Robin have subsequently given birth. Based on their infants' birthdates, it is likely that both females conceived between late November and early January. As we continue to monitor fecal hormone levels, we should be able to determine more precisely gestation length in this species.
18. See Milton (1985b).
19. See Milton (1985b), and Strier (1987c, 1992).
20. See Brownlee (1987).
21. See Harcourt et al. (1981).
22. See Short (1981).
23. See Wrangham (1979), and McFarland Symington (1990).
24. See Tutin (1979).
25. See Strier (1992, in press, b).
26. See de Waal (1986), and F.D.C. Mendes (1990) for discussion of unconventional hierarchical relationships.
27. See Strier (1992, in press, b).
28. See F.D.C. Mendes (1990).
29. See Strier (in press, c).
30. See Strier (in press, c), and A.O. Rímoli (forthcoming thesis). Rímoli conducted a nine month study focusing exclusively on infant and juvenile muriqui development; her work will expand our understanding of this important stage of muriqui social life.
31. Grooming serves both hygienic functions—because it involves the removal of ectoparasites—as well as social functions—because it generally occurs between individuals who have strong social relationships as well. Certainly muriquis have ectoparasites; they frequently scratch themselves, and one population of captured and released muriquis were infested with small flea-like parasites. It is possible that grooming is absent in muriquis because the absence or reduced state of their thumb (probably for suspensory locomotion) inhibits the fine-grain coordination necessary.
32. During his visit in 1988, Frans de Waal observed one female embrace another who was screaming. He suggested that this might have been an example of reconciliation behavior in muriquis. See de Waal (1989).
33. See Strier (1992, in press, b).
34. Despite the complexity of their vocalizations, muriquis can remain very

quiet for long periods of time. During the early days of my research, I sometimes passed directly beneath muriquis without knowing that they were sleeping soundly in tall tree tops. Listening for their calls is one of the ways that we locate the muriquis, but all of us who study muriquis have learned not to rely on vocalizations alone to find the monkeys.

As I have visited other forests to observe other populations of muriquis, I have been struck both by how different some of the vocalizations are, and by how infrequently muriquis in some of the larger forests vocalize. Muriquis may remain quiet when they detect unfamiliar humans in areas where they are still vulnerable to hunters and their silence is their only defense. It is also possible that lower densities or weaker relationships among group members make vocal communication less important in some populations than it appears to be among the muriquis at Montes Claros.

Unraveling what muriquis say when they call to each other, and what listeners understand, as well as differences between the "dialects" of different populations, are intriguing areas of research that have been studied in other primates in an effort to understand their cognitive abilities (Cheney and Seyfarth, 1990), and their species' evolution (e.g. Mitani, 1987). It is clear that many questions about the muriquis' social evolution will only be addressed when we can integrate our long-term behavioral observations with a more detailed understanding of their vocal communication.

35. During my first visit to Montes Claros, the local people told me that when Nishimura had arrived and asked the local farmers where the muriquis were, they took him to the stables. Listening to the horses whinnying back and forth, they told him that he should walk in the forest until he heard horses. Then he would know he had found the muriquis.

36. During my own study in 1983–1984, I distinguished 10 vocalizations by broad acoustic and social context (Strier, 1986). In 1990, F.D.C. Mendes began a systematic study of muriqui vocal communication. His study will permit us to assess whether muriquis understand one another better than we understand them. Focusing exclusively on vocalizations, Mendes can already distinguish more than 20 calls, providing a strong indication that the muriquis' vocal communication, like virtually all other aspects of their lives, is far more complex than originally suspected.

37. See Hauser and Wrangham (1987).

38. See Strier et al. (in press).

39. See Altmann and Altmann (1979).

NOTES TO CHAPTER 7

1. See Strier (1991a).

2. See Chapter 6, especially references to Adriana Rímoli's study on infant and juvenile muriquis.

3. See Strier (1986).

4. See Strier (1986).

5. See Strier (1991a).

6. See Chapter 6 for more details; and Strier (1987c).

7. See Southwick and Smith (1986).

8. See Strier (1990).

9. See Rosenberger and Strier (1989).

10. See Wrangham (1979) for chimpanzees, and Strier (in press, b and c) for muriquis.

11. See Moore (1984), Moore and Ali (1984), and Pusey and Packer (1987) for reviews of inbreeding avoidance and other correlates of dispersal.

12. For review of kin recognition in primates, see Gray (1985:98–101). See also Strier et al. (in press) for discussion of a newly-formed third group at Montes Claros, comprised of both Jaó males and females and Matão female emigrants.

13. All subsequent data on muriqui reproductive parameters come from Strier (1991a).

14. See Strier (1987c).

15. See Western (1979), and Harvey et al. (1987).

16. Collecting fresh fecal material was often quite difficult, because it required that I keep the female I was targeting in constant view. Muriquis often defecate during brief pauses in their travel, and unless I was directly below a female I would not know that she had defecated or where the dung had landed. Resting females were much easier to collect from, provided that they were not in close proximity to any other individuals whose feces might be confused.

Once a fresh fecal sample was located, it was necessary to preserve it immediately so that the steroids would not begin to degenerate. Using tweezers, twigs, and sometimes my fingers, I would transfer the fecal material to a glass vial filled with 10 milliliters of alcohol. The alcohol acted as a short-term presevative until I could freeze the samples in the gas-run freezer at the research house each evening.

Every few days, as the freezer filled up, the samples were transferred on ice to an electric freezer in Sto. Antonio. At the end of my field season, I would fill a large ice-chest with ice and frozen vials, and take the overnight bus to Rio de Janeiro. There I would refill the ice chest with dry ice for the trip back to the United States.

Beginning in 1991, this tedious collection process has been continued by the students who work on the project. The copious quantities of feces they collect means that I often carry up to 70 kilograms of frozen samples on my return trips.

17. See Altmann and Altmann (1979), and Dunbar (1987).

18. See Strier (1991a). Also see Gibbons (1992) for a review of my recent findings, which suggest that plant estrogens may regulate muriqui fertility.

19. See Collias and Southwick (1952), Smith (1977), Dunbar (1980), and Goodall (1983).

20. See Dittus (1977).

21. See Strier et al. (in press).

22. See Strier (in press, c), and Strier et al. (in press).

NOTES TO CHAPTER 8

1. See references in Chapters 6 and 7 to fecal hormone study.

2. Other flagship species include the Mountain Gorilla for Rwanda and the Panda Bear for China; in the Brazilian Atlantic forest, the golden lion tamarin is another important flagship species. See Kleiman (1984), and Dietz et al. (1986) for review of the golden lion tamarin project.

3. See Chapters 2 and 5.

4. In fact, Sr. Bento asked me what I thought a muriqui was worth so that he could provide the prosecution with a minimum penalty. It was a difficult question; how much *is* a muriqui worth? Sr. Bento thought one way to derive a "price" was to estimate how much it would cost to keep a muriqui alive in a zoo. Multiplied over an estimated 30 year lifespan, and the lifespans of the offspring each female victim might have had, resulted in an extraordinary *value* for a muriqui.

5. See Chapter 5.

6. See Janzen (1988); also see J. Cherfas, *New Scientist*, 23 October 1986, pp. 26–27.

7. See Lande (1988), and Lewin (1989) for perspectives on the effects of inbreeding depression.

8. Very few people have the kind of experience necessary for careful capture–release projects. In the United States, if not the world, Dr. Ken Glander of Duke University is the most experienced, with roughly 1,000 successful capture–releases.

9. See J. Brooke, Gold monkeys learn how to live in wild in Brazilian preserve (*The New York Times*, Tuesday, 17 October 1989, C4) for a description of the ways in which captive-born golden lion tamarins were *educated* about the unfamiliar forest into which they were reintroduced.

10. See Chapter 5.

11. See Chapter 7.

12. See Dobson and Lyles (1989).

13. See Strier (1991a).

14. See Griffith et al. (1989) for alternative arguments.

15. The Rio de Janeiro Primate Center was opened by the Rio de Janeiro State Foundation for Environmental Engineering (FEEMA) in 1979 with the aim of establishing breeding programs for the endangered primates of eastern Brazil. It is situated roughly 100 kilometers from the center of Rio de Janeiro, in the foothills of the still-forested Serra dos Órgãos. The Primate Center is most widely known for its stunning success in the captive breeding of the golden lion tamarin, *Leontopithecus rosalia*, and other species of *Leontopithecus* and *Callithrix*, the marmosets.

16. See Pires Vaz (1989); with the more recent arrival of a sexually mature male and the subsequent birth of an infant, the captive breeding program is well on its way and requests for more males have ceased.

17. See Introduction.

18. I was fortunate that the muriquis at this site have such distinct natural markings, and that there was no danger of getting lost in the forest because of its small size. Researchers working with less-distinct looking primates, such as those with similarly dark faces, or in extensive continuous forests where becoming lost is a danger, may have no option other than to mark or radio-track their study subjects.

19. See Chapter 3.

20. Indeed, each muriqui birth is now reported in the city of Caratinga's Saturday news column!

21. See Chapter 5.

22. I saw my first ocelot in the forest at Montes Claros in 1990. Although ocelots were reported to have been numerous in the area in the past, none of the scientists working at the forest since the 1970s had seen one here before. Evidence

that a predator as large as an ocelot still inhabits this forest is yet one more example of the importance of this area for conservation.

23. See references in Chapter 5.

24. Some justification could be made about the greater importance of preserving the muriqui over the brown howler monkey simply on the basis of the fact that the muriqui is an endangered *genus*. The brown howler monkey, by contrast, is an endangered *species*, but there are other species of howler monkeys less at risk in other parts of Central and South America.

25. See Chapter 5.

26. See Chapter 7.

27. See Chapter 6.

28. See Chapter 7; and Strier (in press, c) and Strier et al. (in press).

29. See R. Wright, Banker to the world—and the environment, *The Boston Globe*, 22 August 1988 (business section) for insights into the president of the World Bank and his commitments to global ecology.

30. See Sun (1988).

31. See G. de Lama, Brazil resists campaign to save rain forests, *Chicago Tribune*, 12 March 1989 (front page article).

References

Aguirre, A.C. (1971). *O mono Brachyteles arachnoides.* Academia Brasileira de Ciencias, Rio de Janeiro.

Altmann, J. (1974). Observational study of behavior: sampling methods. *Behaviour*, 48:1–41.

Altmann, J. (1980). *Baboon mothers and infants.* Harvard University Press, Cambridge, MA.

Altmann, S.A. and Altmann, J. (1979). Demographic constraints on behavior and social organization. In T.H. Bernstein and E.O. Smith, eds., *Primate ecology and human evolution*, pp. 47–63. Garland, New York.

Alves, M.C. (1986). Novas localizações do mono carvoeiro, *Brachyteles arachnoides* (Cebidea, Primates) e situação do parque nacional do Caparaó. In M. Thiago de Mello, ed., *A primatologia no Brasil—2*, pp. 367–368. Sociedade Brasileira de Primatologia, Brasilia.

Bauchop, T. (1978). Digestion of leaves in vertebrate arboreal folivores. In G.G. Montgomery, ed., *The ecology of arboreal folivores*, pp. 193–204. Smithsonian Institution Press, Washington, D.C.

Bradbury, J.W. and Vehrencamp, S.L. (1977). Social organization and foraging in emballonurid bats, III: mating systems. *Behavioral Ecology and Sociobiology*, 2:1–17.

Brownlee, S. (1987). These are real swinging primates. *Discover*, 4:66–77.

Camara Cascudo, L. da (1985). *Superstição no Brasil.* Editora da Universidade de São Paulo, São Paulo.

Cant, J.G.H. (1977). *Ecology, locomotion, and social organization of spider monkeys (Ateles geoffroyi).* Ph.D. dissertation, University of California, Davis.

Cant, J.G.H. (1986). Locomotion and feeding postures of spider and howling monkeys: field study and evolutionary interpretation. *Folio Primatologica*, 46:1–14.

Cheney, D.L. and Seyfarth, R.M. (1990). *How monkeys see the world.* University of Chicago Press, Chicago.

Chivers, D.J., Wood, B.A., and Bilsborough, A. (1984). *Food acquisition and processing in primates.* Plenum Press, New York.

Ciochon, R.L. and Chiarelli, A.B. (1980). Paleobiogeographic perspectives on the origin of the platyrrhini. In R.L. Ciochon and A.B. Chiarelli, eds., *Evolutionary biology of the New World monkeys and continental drift*, pp. 459–493. Plenum Press, New York.

Clutton-Brock, T.H. (1977). Some aspects of intraspecific variation in feeding and ranging behaviour in primates. In T.H. Clutton-Brock, ed., *Primate ecology*, pp. 539–556. Academic Press, New York.

Clutton-Brock, T.H. and Harvey, P.H. (1977). Species differences in feeding and ranging behaviour in primates. In T.H. Clutton-Brock, ed., *Primate ecology*, pp. 557–584. Academic Press, New York.

Clutton-Brock, T.H. and Harvey, P.H. (1979). Home range size, population density and phylogeny in primates. In I.S. Bernstein and E.O. Smith, eds., *Primate ecology and human origins*, pp. 201–214. Gartland STPM Press, New York.

Clutton-Brock, T.H. and Harvey, P.H. (1984). Comparative approaches to investigating adaptation. In J.R. Krebs and N.B. Davies, eds., *Behavioural ecology*, pp. 7–29. Blackwell Scientific Publications, Boston.

Clutton-Brock, T.H., Harvey, P.H., and Rudder, B. (1977). Sexual dimorphism, socionomic sex ratio and body weight in primates. *Nature*, 269:797–800.

Coimbra-Filho, A.F. (1972). Mamíferos ameaçados de extinção no Brasil. In *Espécies da fauna Brasileira ameaçadas de extinção*, pp. 13–98. Brazilian Academy of Sciences, Rio de Janeiro.

Coimbra-Filho, A.F. (1984). Situação de fauna na floresta Atlântica. *Boletim FBCN*, 19:89–110.

Collias, N. and Southwick, C. (1952). A field study of population density and social organization in howling monkeys. *Proceedings of the American Philosophical Society*, 96:143–156.

Conroy, G.C. (1990). *Primate evolution.* W.W. Norton & Company, New York.

Darwin, C. (1871). *The descent of man, and selection in relation to sex.* Appleton, New York.

Demment, M.W. (1983). Feeding ecology and the evolution of body size of baboons. *African Journal of Ecology*, 21:219–233.

Dietz, J.M., Coimbra-Filho, A.F., and Pessamílio, D.M. (1986). Projeto mico-leão. I. Um modelo para a conservação de espécie ameaçada de extinção. In M. Thiago de Mello, ed., *A primatologia no Brasil—2*, pp. 217–222. Sociedade Brasileira de Primatologia, Brasilia.

Dittus, W. (1977). The social regulations of population density and age–sex distribution in the toque monkey. *Behaviour*, 63:281–321.

Dobson, A.P. and Lyles, A.M. (1989). The population dynamics and conservation of primate populations. *Conservation Biology*, 3:362–380.

Dunbar, R.I.M. (1980). Demographic and life history variables of a population of gelada baboons (*Theropithecus gelada*). *Journal of Animal Ecology*, 49:485–506.

Dunbar, R.I.M. (1987). Demography and reproduction. In B.B. Smuts, D.L. Cheney, R.M. Seyfarth, R.W. Wrangham, and T.T. Struhsaker, eds., *Primate societies*, pp. 240–249. University of Chicago Press, Chicago.

Eisenberg, J.F., Muckenhirn, N.A., and Rudran, R. (1972). The relation between ecology and social structure in primates. *Science*, 176:863–874.

Emlen, S.T. and Oring, L.W. (1977). Ecology, sexual selection, and the evolution of mating systems. *Science*, 197:215–223.

Emmons, L.H. (1990). *Neotropical rainforest mammals.* University of Chicago Press, Chicago.

Ferrari, S.F. (1988). *The behaviour and ecology of the buffy-headed marmoset, Callithrix flaviceps (O. Thomas, 1903).* Ph.D. dissertation, University College London.

Ferrari, S.F. and Strier, K.B. (1992). Exploitation of *Mabea fistulifera* nectar by marmosets (*Callithrix flaviceps*) and muriquis (*Brachyteles arachnoides*). *Journal of Tropical Ecology*, 8:225–239.

Fisher, R.A. and Yates, F. (1957). *Statistical tables for biological, agricultural and medical research.* Hafner Publishing Company Inc., New York.

Fleagle, J.G. (1989). *Primate adaptation and evolution.* Academic Press, New York.

Fonseca, G.A.B. (1983). *The role of deforestation and private reserves in the conservation of the woolly spider monkey (Brachyteles arachnoides).* MA thesis, University of Florida, Gainesville.

Fonseca, G.A.B. (1985). The vanishing Brazilian Atlantic forest. *Biological Conservation,* 34:17–34.

Gaulin, S.J.C. (1979). A Jarman/Bell model of primate feeding niches. *Human Ecology,* 7:1–20.

Gibbons, A. (1992). Plants of the apes. *Science,* 255:921.

Goodall, J. (1983). Population dynamics during a 15-year period in one community of free-living chimpanzees in the Gombe National Park, Tanzania. *Zeitschrift für Tierpsychologie,* 61:1–60.

Gray, J.P. (1985). *Primate sociobiology.* HRAF Press, New Haven, Connecticut.

Griffith, B., Michael Scott, J., Carpenter, J.W., and Reed, C. (1989). Translocation as a species conservation tool: status and strategy. *Science,* 245:477–480.

Haffer, J. (1969). Speciation in Amazonian forest birds. *Science,* 165:131–137.

Haffer, J. (1974). *Avian speciation in tropical South America.* Publication No. 14, Nuttal Ornithological Club, Cambridge, MA.

Harcourt, A.H., Harvey, P.H., Larson, S.G.P., and Short, R.V. (1981). Testis weight, body weight and breeding system in primates. *Nature,* 293:55–57.

Harvey, P.H., Martin, R.D., and Clutton-Brock, T.H. (1987). Life histories in comparative perspective. In B.B. Smuts, D.L. Cheney, R.M. Seyfarth, R.W. Wrangham, and T.T. Struhsaker, eds., *Primate societies,* pp. 181–196, University of Chicago Press, Chicago.

Hatton, J.C., Smart, N.O.E., and Thomson, K. (1983). *An ecological study of the Fazenda Montes Claros forest, Minas Gerais, Brazil.* Interim report, Department of Botany and Microbiology, University College London.

Hauser, M.D. and Wrangham, R.W. (1987). Manipulation of food calls in captive chimpanzees. *Folia Primatologica,* 48:207–210.

Hill, W.C.O. (1962). *Primates: comparative anatomy and taxonomy, volume 5: Cebidae part (b).* Wiley–Interscience, New York.

Huffman, M.A. and Seifu, M. (1989). Observations on the illness and consumption of a possibly medicinal plant *Vernonia amygdalina* (Del.), by a wild chimpanzee in the Mahale Mountains National Park, Tanzania. *Primates,* 30:51–63.

IUCN—International Union for Conservation of Nature and Natural Resources (1982). *The IUCN Mammal Red Data Book, Part 1.* IUCN, Switzerland.

Jackson, J.F. (1978). Differentiation in the genera *Enyalius* and *Strobilurus* (Iguanidae): implications for Pleistocene climate changes in eastern Brazil. *Arquivos de Zoologia (São Paulo),* 30:1–79.

Janzen, D.H. (1966). Coevolution of mutualism between ants and acacias in Central America. *Evolution,* 20:249–275.

Janzen, D.H. (1988). Tropical ecological and biocultural restoration. *Science,* 239:243–244.

Jarman, P.J. (1974). The social organisation of antelope in relation to their ecology. *Behaviour,* 48:215–267.

Kay, R.F., Plavcan, J.M., Glander, K.E., and Wright, P.C. (1988). Sexual selection and canine dimorphism in New World monkeys. *American Journal of Physical Anthropology,* 77:385–397.

Kinzey, W.G. (1982). Distribution of primates and forest refuges. In G.T. Prance, ed., *Biological diversification in the tropics*, pp. 455–482. Columbia University Press, New York.

Kleiman, D.G. (1984). The behavior and conservation of the golden lion tamarin, *Leontopithecus r. rosalia*. In M. Thiago de Mello, ed., *A primatologia no Brasil—1*, pp. 35–53. Sociedade Brasileira de Primatologia, Brasilia.

Kricher, J.C. (1989). *A neotropical companion*. Princeton University Press, Princeton.

Lande, R. (1988). Genetics and demography in biological conservation. *Science*, 241:1455–1460.

Leighton, M. and Leighton, D.R. (1982). The relationship of size of feeding aggregate to size of food patch: howler monkeys (*Alouatta palliata*) feeding in *Trichilia cipo* fruit trees on Barro Colorado Island. *Biotropica*, 14:81–90.

Lemos de Sá, R.M. (1988). *Situação de uma população de mono-carvoeiro, Brachyteles arachnoides, em fragmento de mata Atlântica (M.G.), e implicações para sua conservação.* MA thesis, Universidade de Brasilia, Brasilia.

Lemos de Sá, R.M. and Glander, K. (in press). Morphometrics of the woolly spider monkey, or muriqui (*Brachyteles arachnoides*, E. Geoffroy 1806). *American Journal of Primatology*.

Lemos de Sá, R.M. and Strier, K.B. (in press). Comparative forest structure and habitat choice in muriquis. *Biotropica*.

Leutenegger, W. and Kelly, J.T. (1977). Relationship of sexual dimorphism in canine size and body size to social behavioral and ecological correlates in anthropoid primates. *Primates*, 18:117–136.

Lewin, R. (1984). DNA reveals surprises in human family tree. *Science*, 225:1179–1182.

Lewin, R. (1988a). Conflict over DNA clock results. *Science*, 241:1598–1600.

Lewin, R. (1988b). DNA clock conflict continues. *Science*, 241:1756–1759.

Lewin, R. (1989). Inbreeding costs swamp benefits. *Science*, 243:482.

Lindstedt, S.L. and Boyce, M.S. (1985). Seasonality, fasting endurance and body size in mammals. *American Naturalist*, 125:873–878.

Marks, J. (1991). What's old and new in molecular phylogenetics. *American Journal of Physical Anthropology*, 85:207–219.

McFarland Symington, M. (1990). Fission–fusion social organization in *Ateles* and *Pan*. *International Journal of Primatology*, 11:47–61.

McNab, B.K. (1978). Energetics of arboreal folivores: physiological problems and ecological consequences of feeding on an ubiquitous food supply. In G.G. Montgomery, ed., *The ecology of arboreal folivores*, pp. 153–162. Smithsonian Institution Press, Washington, D.C.

Mendes, F.D.C. (1990). *Affiliação e hierarquia no muriqui: o grupo Matão de Caratinga.* MA thesis, Universidade de São Paulo, São Paulo.

Mendes, S.L. (1985). *Uso de espaço, padrões de atividades diárias e organização social de Alouatta fusca (Primatas, Cebidae) em Caratinga, M.G.* MA thesis, Universidade de Brasilia, Brasilia.

Mendes, S.L. (1989). Estudo ecologico de *Alouatta fusca* (Primates: Cebidae) na Estação Biologica de Caratinga, M.G. *Revista Nordestina Biologica*, 6:71–104.

Milton, K. (1980). *The foraging strategy of howler monkeys*. Columbia University Press, New York.

Milton, K. (1984a). The role of food-processing factors in primate food choice. In P.S. Rodman and J.G.H. Cant, eds., pp. 249–279. *Adaptations for foraging in nonhuman primates*. Columbia University Press, New York.

Milton, K. (1984b). Habitat, diet, and activity patterns of free-ranging woolly spider monkeys (*Brachyteles arachnoides* E. Geoffroy 1806). *International Journal of Primatology*, 5:491–514.

Milton, K. (1985a). Multimale mating and absence of canine tooth dimorphism in woolly spider monkeys (*Brachyteles arachnoides*). *American Journal of Physical Anthropology*, 68:519–523.

Milton, K. (1985b). Mating patterns of woolly spider monkeys, *Brachyteles arachnoides*: implications for female choice. *Behavioral Ecology and Sociobiology*, 17:53–59.

Milton, K. (1985c). Urine washing behavior in the woolly spider monkey (*Brachyteles arachnoides*). *Zeitschrift für Tierpsychologie*, 67:154–160.

Milton, K. and de Lucca, C. (1984). Population estimate for *Brachyteles* at Fazenda Barreiro Rico, São Paulo State, Brazil. *IUCN/SSC Primate Specialist Group Newsletter*, 4:27–28.

Mitani, J. (1987). Species discrimination of male song in gibbons. *American Journal of Primatology*, 13:413–423.

Mittermeier, R.A., Coimbra-Filho, A.F., Constable, I.D., Rylands, A.B., and Valle, C. (1982). Conservation of primates in the Atlantic forest region of eastern Brazil. *International Zoo Yearbook*, 22:2–17.

Mittermeier, R.A., Valle, C.M.C., Alves, M.C., Santos, I.B., Pinto, C.A.M., Strier, K.B., Young, A.L., Veado, E.M., Constable, I.D., Paccagnella, S.G., and Lemos de Sá, R.M. (1987). Current distribution of the muriqui in the Atlantic forest region of eastern Brazil. *Primate Conservation*, 8:143–149.

Miyamoto, M., Slighton, J., and Goodman, M. (1987). Phylogenetic relations of humans and African apes from DNA sequences in the pseudo-eta globin region. *Science*, 238:369–373.

Moore, J. (1984). Female transfer in primates. *International Journal of Primatology*, 5:537–589.

Moore, J. and Ali, R. (1984). Are dispersal and inbreeding avoidance related? *Animal Behaviour*, 32:94–112.

Mori, A. (1979). Analysis of population changes by measurements of body weight in the Koshima troop of Japanese monkeys. *Primates*, 20:371–397.

Mori, S.A., Boom, B.M., and Prance, G.T. (1981). Distribution patterns and conservation of eastern Brazilian coastal forest tree species. *Brittonia*, 33:233–245.

Moynihan, M. (1967). *The New World primates*. Princeton University Press, Princeton, New Jersey.

Müller, P. (1973). *The dispersal centres of terrestrial vertebrates in the Neotropical realm*. W. Junk Publishers, The Hague.

Napier, P.H. (1976). *Catalogue of primates in the British Museum (Natural History). Part I: Families Callitrichidae and Cebidae*. British Museum (Natural History), London.

National Academy Press (1981). *Techniques for the study of primate population ecology*. National Academy Press, Washington, DC.

Nishimura, A. (1979). In search of woolly spider monkey. *Kyoto University Overseas Research Reports of New World Monkeys*, pp. 21–37.

Paccagnella, S.G. (1986). *Relatório sobre o censo da população de monos-carvoeiros do parque estadual de "Carlos Botelho"*. Unpublished.

Parra, R. (1978). Comparison of foregut and hindgut fermentation in herbivores. In G.G. Montgomery, ed., *The ecology of arboreal folivores*, pp. 205–230. Smithsonian Institution Press, Washington, D.C.

Parsons, P.E. and Taylor, C.R. (1977). Energetics of brachiation versus walking: a comparison of a suspended and an inverted pendulum mechanism. *Physiological Zoology*, 50:182–188.

Phillips-Conroy, J. (1986). Baboons, diet, and disease: food plant selection and schistosomiasis. In D.M. Taub and F.A. King, eds., *Current perspectives in primate social dynamics*, pp. 287–304. Van Nostrand, New York.

Pinto, L.P.S., Costa, C.M.R., Strier, K.B., and Fonseca, G.A.B. (submitted). Habitats, density, and group size of primates in the Reserva Biologica Augusto Ruschi (Nova Lombardia), Santa Teresa, Brazil. *Folia Primatologica.*

Pires Vaz, D. (1989). Mundo animal: os macacos ameaçados da mata Atlântica; Centro de primatologia—a ciência a serviço da natureza. *Revista Geográfica Universal*, 180:60–75.

Pusey, A.E. and Packer, C. (1987). Dispersal and philopatry. In B.B. Smuts, D.L. Cheney, R.M. Seyfarth, R.W. Wrangham, and T.T. Struhsaker, eds., *Primate societies*, pp. 250–266. University of Chicago Press, Chicago.

Richard, A. (1985). *Primates in nature.* Freeman, New York.

Rizzini, C.T. and Coimbra-Filho, A.F. (1988). *Ecossistemas brasileiros.* Enge-Rio/Editora Index, Rio de Janeiro.

Rosenberger, A.L. (1981). Systematics: the higher taxa. In A.F. Coimbra-Filho and R.A. Mittermeier, eds., *Ecology and behavior of neotropical primates, volume 1*, pp. 9–109. Academia Brasileira de Ciências, Rio de Janeiro.

Rosenberger, A.L. and Strier, K.B. (1989). Adaptive radiation in the ateline primates. *Journal of Human Evolution*, 18:717–750.

Santos, I.B., Mittermeier, R.A., Rylands, A.B., and Valle, C.M.C. (1987). The distribution and conservation status of primates in southern Bahia, Brazil. *Primate Conservation*, 8:126–142.

Schaik, C.P. van and Hooff, J.A.R.A.M. van (1983). On the ultimate causes of primate social systems. *Behaviour*, 85:91–117.

Schultes, R.E. and Raffauf, R.F. (1990). *The healing forest.* Dioscorides Press, Portland, Oregon.

Short, R.V. (1981). *Sexual selection in man and the great apes.* Academic Press, New York.

Sibley, C.G. and Ahlquist, J. (1984). The phylogeny of the hominoid primates, as indicated by DNA–DNA hybridization. *Journal of Molecular Evolution*, 20:2–15.

Small, M.F. (1981). Body fat, rank, and nutritional status in a captive group of Rhesus macaques. *International Journal of Primatology*, 2:91–95.

Smith, C.C. (1977). Feeding behaviour and social organization in howling monkeys. In T.H. Clutton-Brock, ed., *Primate ecology*, pp. 97–126. Academic Press, New York.

Smuts, B.B. (1985). *Sex and friendship in baboons.* Aldine Publishing Company, New York.

Smuts, B.B. (1987). Gender, aggression, and influence. In B.B. Smuts, D.L. Cheney, R.M. Seyfarth, R.W. Wrangham, and T.T. Struhsaker, eds., *Primate societies*, pp. 400–412. University of Chicago Press, Chicago.

Southwick, C.H. and Smith, R.B. (1986). The growth of primate field studies. In G. Mitchell and J. Erwin, eds., *Comparative primate biology, volume 2a: Behavior, conservation, and ecology*, pp. 73–91. Alan R. Liss, New York.

Stallings, J.R. (1988). Small mammal inventories in an eastern Brazilian park. *Bulletin of the Florida State Museum Biological Sciences*, 34:153–200.

Strier, K.B. (1986). *The behavior and ecology of the woolly spider monkey, or muriqui* (*Brachyteles arachnoides, E. Geoffroy 1806*). Ph.D. dissertation, Harvard University, Cambridge, MA.

Strier, K.B. (1987a). Activity budgets of woolly spider monkeys, or muriquis. *American Journal of Primatology*, 13:385–395.

Strier, K.B. (1987b). Ranging behavior of woolly spider monkeys. *International Journal of Primatology*, 8:575–591.

Strier, K.B. (1987c). Reprodução de *Brachyteles arachnoides.* In M. Thiago de Mello, ed., *A primatologia no Brasil—2*, pp. 163–175. Sociedade Brasileira de Primatologia, Brasilia.

Strier, K.B. (1989). Effects of patch size on feeding associations in muriquis (*Brachyteles arachnoides*). *Folia Primatologica*, 52:70–77.

Strier, K.B. (1990). New World primates, new frontiers: insights from the woolly spider monkey, or muriqui (*Brachyteles arachnoides*). *International Journal of Primatology*, 11:7–19.

Strier, K.B. (1991a). Demography and conservation in an endangered primate, *Brachyteles arachnoides. Conservation Biology*, 5:214–218.

Strier, K.B. (1991b). Diet in one group of woolly spider monkeys, or muriquis (*Brachyteles arachnoides*). *American Journal of Primatology*, 23:113–126.

Strier, K.B. (1992). Causes and consequences of nonaggression in woolly spider monkeys. In J. Silverberg and J. Patrick Gray, eds., *Aggression and peacefulness in humans and other primates*, pp. 100–116. Oxford University Press, New York.

Strier, K.B. (in press, a). Atelinae adaptations: behavioral strategies and ecological constraints. *American Journal of Physical Anthropology.*

Strier, K.B. (in press, b). Subtle cues of social relations in male muriqui monkeys (*Brachyteles arachnoides*). In W.G. Kinzey, ed., *New World primates: ecology, evolution, and behavior.* Aldine de Gruyter, New York.

Strier, K.B. (in press, c). Growing up in a patrifocal society: sex differences in the spatial relations of immature muriquis (*Brachyteles arachnoides*). In M.E. Pereira and L.A. Fairbanks, eds., *Juveniles: comparative socioecology.* Oxford University Press, New York.

Strier, K.B. and Stuart, M.D. (1992). Intestinal parasites in the muriqui (*Brachyteles arachnoides*): population variability, ecology, and conservation. *American Journal of Physical Anthropology*, 14:158–159.

Strier, K.B., Mendes, F.D.C., Rímoli, J., and Rímoli, A.O. (in press). Demography and social structure of one group of muriquis (*Brachyteles arachnoides*). *International Journal of Primatology.*

Stuart, M.D., Strier, K.B., and Pierberg, S.M. (in press). A coprological survey of parasites of wild muriquis, *Brachyteles arachnoides*, and brown howling monkeys, *Alouatta fusca. Journal of the Helminthological Society of Washington.*

Sun, M. (1988). Costa Rica's campaign for conservation. *Science*, 239:1366–1369.

Symington, M.M. (1988). Food competition and foraging party size in the black spider monkey (*Ateles paniscus chamek*). *Behaviour*, 105:117–134.

Terborgh, J. (1985). *Five New World primates.* Princeton University Press, Princeton, New Jersey.

Torres de Assumpção, C. (1981). *Cebus apella* and *Brachyteles arachnoides* (Cebidae) as potential pollinators of *Mabea fistulifera* (Euphorbiaceae). *Journal of Mammalogy*, 62:386–388.

Torres de Assumpção, C. (1983a). *An ecological study of the primates of southeastern Brazil, with a reappraisal of Cebus apella races.* Ph.D. dissertation, University of Edinburgh, Edinburgh.

Torres de Assumpção, C. (1983b). Ecological and behavioural information on *Brachyteles arachnoides. Primates,* 24:584–593.

Torres de Assumpção, C., Leitão-Filho, H.F., and Cesar, O. (1982). Descrição das matas da Fazenda Barreiro Rico, estado de São Paulo. *Revista Brasileira Botanica,* 5:53–66.

Trivers, R.L. (1972). Parental investment and sexual selection. In B. Campbell, ed., *Sexual selection and the descent of man.* Aldine Publishing Company, Chicago.

Tutin, C.E.G. (1979). Mating patterns and reproductive strategies in a community of wild chimpanzees (*Pan troglodytes schweinfurthii*). *Behavioral Ecology and Sociobiology,* 6:29–38.

Valle, C.M.C., Santos, I.B., Alves, M.C., Pinto, C.A., and Mittermeier, R. (1984). Algumas observações preliminares sobre o comportamento do mono (*Brachyteles arachnoides*) em ambiente natural (Fazenda Montes Claros, município de Caratinga, Minas Gerais, Brasil). In M. Thiago de Mello, ed., *A primatologia no Brasil—1,* pp. 271–283. Sociedade Brasileira de Primatologia, Brasilia.

Vieira, C.C. (1944). Os símios do estado de São Paulo. *Papeis Avulsos de Zoologia (São Paulo),* 4:1–31.

Vieira, C.C. (1955). Lista remissiva dos mamíferos do Brasil. *Arquivos de Zoologia (São Paulo),* 8:341–474.

de Waal, F. (1986). Integration of dominance and social bonding in primates. *Quarterly Review of Biology,* 61:459–479.

de Waal, F. (1989). *Peacemaking among primates.* Harvard University Press, Cambridge, MA.

Western, D. (1979). Size, life history and ecology in mammals. *African Journal of Ecology,* 17:185–204.

White, F.J. and Wrangham, R.W. (1988). Feeding competition and patch size in the chimpanzee species *Pan paniscus* and *Pan troglodytes. Behaviour,* 105:148–164.

Wied-Neuwied, Prinz M. von (1958). *Viagem ao Brasil.* Translated by E. Sussekind de Mendonça and F. Poppe de Figueiredo. Cia. Editora Nacional, São Paulo.

Wolfheim, J.H. (1983). *Primates of the world.* University of Washington Press, Seattle, Washington.

Wrangham, R.W. (1979). On the evolution of ape social systems. *Social Science Information,* 18:335–386.

Wrangham, R.W. (1980). An ecological model of female-bonded primate groups. *Behaviour,* 75:262–299.

Wrangham, R.W. and Nishida, T. (1983). *Aspilia* spp. leaves: a puzzle in the feeding behavior of wild chimpanzees. *Primates,* 24:276–282.

Zingeser, M.R. (1973). Dentition of *Brachyteles arachnoides* with reference to Alouattine and Atelinine affinities. *Folia Primatologica,* 20:351–390.

Index

Abdalla, Sr. Feliciano Miguel 13, 18, 46, 102–3
Activity cycle 50
Adenocalymna 53
Adolescent females 87, 88, 90
Adolescent males 87
Affection 76
Aggression 67, 68, 70, 96, 110
Agility 3
Alberto, Carlos 25, 35, 44, 45
Alkaloids 53
Alouatta, see Howler monkeys
Alouatta fusca 5, 20
Alvaro Coutinho Aguirre 13
Alves, Cristina 43
Amazon 5–6, 9, 112
Amino acids 53
Antibacterial compounds 56
Antiparasitic agents 56
Apuleia 53, 60
Ateles, see Spider monkeys
Atelinae 5
Atlantic forest 9, 112
Augusto Ruschi Biological Reserve 62, 101

Baboons 56, 70, 89
Bahia refuge 10, 11, 13
Barking 80
Behavioral patterns 31, 83
Behavioral sampling 30–1
Belo Horizonte 42, 43
Biodiversitas 13
Biological Station of Caratinga 13
Birth
  intervals 92
  rate 105
  season 92–3
Body size 4, 70
Botoccudo Indians 12
Brachyteles arachnoides 5
Brachyteles arachnoides arachnoides 10
Brachyteles arachnoides hypoxanthus 10
Brazilian Foundation for the Conservation of Nature 13, 35
Brown howler monkey 5, 20

Callithrix flaviceps 20
Calls 79 (*see also* Vocalizations)
Camping behavior 59
Canine tooth size 4, 70
Canopy volume 32–3
Captive breeding colonies 105–6
Capuchin monkeys 20, 23, 99
Caratinga 18, 42
Carlos Botelho forest, *see* Carlos Botelho State Park
Carlos Botelho State Park 13, 62, 101, 109
Cebus apella nigritus 20
Cecropia 55, 56
Charcoal monkey 4
Chases 68
Chemical signals 69
Chimpanzees 16, 55, 73, 95
Chuckles 78
Coffee plantations 18–19
Coimbra-Filho, Dr. Adelmar 18, 106
Competition 89
Conception 90
Conservation
  agencies 102
  concerns 98–112
  status 17
Costa Rica 111
Coutinho, Paulo 49

Darwin, Charles 70
Deaths 93–4
Debt-for-nature swaps 111
Dependence–independence transition 86
Diet, *see* Feeding habits; Foods
Disappearances 93–4
Diseases 94, 104

Dispersal patterns 88, 89, 110
Dominance relationships 67, 74
Drinking habits 51

Ecological sampling 31–4
Ecotourism 102, 108
Ejaculation/ejaculate plugs 71
Embraces 69, 76, 77
Emigrations/immigrations 87, 88, 93
Endangered species 5, 8, 107
Endemism 9
Energy requirements 4, 52–3, 64
Environmental impact surveys 111
Espírito Santo 13

Facial characteristics 38
Facial coloration 11
Facial markings 49
Facial pigmentation 27, 38
Falling 57, 64, 67, 68
Fazenda Barreiro Rico 60, 69
Fazenda Corrégo de Areia 13
Fazenda Esmeralda 13, 61, 103, 106
Fazenda Montes Claros 13, 14, 18–22, 35, 51, 53, 60, 62, 89, 95, 99–109
Fecal samples 71–2, 90, 92
Feces examination 56
Feeding habits 3, 16, 17, 28–30, 32, 50–65, 79, 92, 104 (*see also* Foods)
Feeding tree focal samples 31
Feliciano, Sr., *see* Abdalla, Sr. Feliciano Miguel
Female behavioral strategies 15
Female-biased dispersal 88, 89
Female-bonded groups 15–17, 89
Female-bonded species 29, 65
Ferreira, Nadir 43, 46
Fire 98, 99
Flagship species 99
Flower production 52–3
Flowering cycles 59
Focal animal samples 31
da Fonseca, Gustavo 36
Food
  distribution 110
  patches 32–3
  resources 31–2, 109
  species 108
Foods 16, 17, 51–64, 60
  flowers 52–3, 64
  fruits 51–3, 56–9, 64
  growth 16
  leaves 53–7, 64
  medicinal properties 56
  new species 43
  reserves 16
  seeds 51, 53, 54
  subsistence 16
  types 28–30
Foraging
  activities 16, 50
  behavior 79
Forest
  clearing 100
  destruction 9–14
  fire 98–9
  fragmentation 109
  structure 60–1
Fur color 27

Gastrointestinal tract 5
Genitalia 4, 69
Gestation period 63, 90, 92
Gomes, Jairo 46
Gorillas 16, 89
Grappling 86
Group size 82, 93
Group solidarity 79
Grouping patterns, 95
de Gusmão Câmara, Ibsen 18

Habitat types 19, 60
Haplorhine primates 6
Home ranges 59, 95
Hormonal cycles 90, 91
Hostile behavior 66
Howler monkeys 5, 20, 23, 50, 56, 58, 89, 109–110
Hunting 94, 100, 101 (*see also* Poachers and poaching)
Hypotheses, *see* Research design

Immigrations/emigrations 87, 88, 93 (*see also* Dispersal patterns)
Inbreeding 89, 103, 105
Infant behavior patterns 83
Infant–mother relationship 83–4
Inspections 69–70
Intergroup encounters 44–6, 62–4, 68, 80, 81, 88
International Union for the Conservation of Nature (IUCN) Red Data book 8
Interspecific relationships 58, 109–10
Intestinal tract 4
Ipanema 43
Iron extraction 12

Jaguars 109
Jaó group
  behavior of 81
  changes 95, 96
  habitat 20
  intergroup encounters 46, 62–4, 68, 80, 88
  migrations 45, 87, 95
  observations 44
Jaws 3
Jequitiba 25, 26
Juvenile behavior patterns 74, 84–7

Kinship patterns 110 (*see also* Dispersal patterns)

Lactation 63
Lagothrix, *see* Woolly monkeys
Lemos de Sá, Rosa 36, 61
Life histories 82–97
Logging/lumber 18, 19, 21, 101–2

Mabea 33, 40, 53, 59
Macaques 70, 89
Male behavioral strategies 15
Male dispersal 89
Male philopatry 88, 89
Mapping 22
Marmoset 20
Matão group
  emigration 87
  habitat 20–1
  immigrations 87–8, 93
  inbreeding 89
  intergroup encounters 46, 62–4, 68, 81
  social organization 28
Mating behavior 66–81, 89–90
Matrilineal kinship 89
Medicinal properties of foods 56
Mendes, Dida 47, 49
Mendes, Sergio 36
Milton, Katharine 60, 61, 73
Minas Gerais 12, 13, 106
Mittermeier, Russell 17–18, 23, 25, 43–5, 107
Model development 15–17
Molars 3
Mono carvoeiro 4
Morphology, *see* Physical traits
Mother–infant relationship 83–4
Mucuna 54
Muriquis
  characteristics of 3
  early observations 37
  effect of forest destruction 13
  see also Jaó group; Matão group
Myrtaceae (myrtle) berries 52, 59, 63, 64

Naming system 49
Nearest neighbors 30 (*see also* Spatial relationships)
Neighs 26, 39, 79, 80, 85, 88
Neri, Fernanda 49
New World monkeys 6, 7
Newborns 83
Nishimura, Akisato 22
Nonhuman primates 6–7

Ocelots 109
Old World monkeys 6, 7, 16, 89
Ovulation 69

Parasites 56, 104, 109
Patrilineal kinship 73 (*see also* Male philopatry)
Paulista refuge 10, 11
Phenological data 32
Pheromonal signals 68
Physical traits 3
Plant
  defences 56
  samples 33
  species 59
Play behavior 86–7
Plotless quadrant sample 61
Poachers and poaching 62, 101–2
Popularity 74
Population surveys 5, 13, 60, 95
Portuguese exploration 12
Predators 109
Preferred associates 76 (*see also* Spatial relationships)
Pregnancies 63, 90–2
Prehensile tail 3, 32
Primatology 7
Protein from leaves 53
Proximity, *see* Spatial relationships

Rainfall 51
Ranging behavior 44–5, 57–9 (*see also* Traveling)
Red colobus monkeys 89
Reforestation 103, 109

Refuges 9, 10
Regenerating habitats 60
Reproduction 90–2
Reproductive state 69
Research
  design 27–34
  methodology 30
Rímoli, Adriana 47
Rímoli, Zé 47
Rio das Contas 10
Rio de Janeiro 13
Rio de Janeiro Primate Center 105
Rio Doce refuge 10
Rio Doce State Park 101
Rio Doce Valley 12
Rio Pardo 10

Sampling methods 27
Santos, Ilmar 22
São Paulo 9, 12, 13
Sarney, José 111
Scan samples 30–1
Scent 26
Scientific presence 107–8
Screaming 96
Screeching 85
Scrub habitats 60
Seasonal changes in temperature 50
Seasonal cycles 51
Secondary habitats 60
Secondary plant compounds, *see* Plant defences
Seed dispersal 54, 55
Serra do Mar 9
Sex differences 4, 75
Sexual behavior 66–81
Sexual interactions 69–70
Sexual monomorphism 67, 70, 71 (*see also* Body size)
Sexual selection 70
Simonésia 13
Sleeping area 50
Sleeping behavior 79
Smell 26
Social behavior 17, 64
Social bonds 85
Social groups 29
Social interactions 29, 31
Social organization 15, 17, 28, 95, 110
Social pressures 63
Socialization 85
Spatial relationships 30, 74, 75, 78
Sperm competition 73, 74
Spider monkeys 5, 89, 90, 95
Spondius dulci 59
Squirrel monkeys 89
Stuart, Dr. Michael 56
subadulthood 74, 79, 87 (*see also* Adolescent females; Adolescent males)
Sugar cane plantations 12, 44
Supplants 68–9
Suspensory locomotion 3, 57

Tail, *see* Prehensile tail
Tannins 53
Temperature effects on behavior 50
Testes 72
Tolerance of males toward females and one another 66, 74
Toxic compounds 54
Trails 22
Translocations 103–5
Traps 101
Traveling
  behavior 79–80 (*see also* Ranging behavior)
  distances 56, 57, 58
  time 50
  see also Suspensory locomotion
Trumpet vine 53
Twins 93
Twittering 66, 70, 71

Urine washing 69

Valle, Celio 18, 41, 43, 47, 107
Veado, Eduardo 41–2
Vegetation plots 31–2, 40, 53
Vocalizations 46, 66, 78, 80

Warbles 78
Weaning 85–6, 92
Weight 3
Whines 85
Wied-Neuwied, Prince Maximilian 12
Woolly monkeys 5, 89
World Wildlife Fund 17
Wrangham, Richard 15, 17

Young, Andy 23, 43–5

Ziegler, Dr. Toni 90